Lion *(Panthera leo)*

Relaxed Elephant herd surrounding vehicle to reach their favourite trees

CPZ 426 EC

African Buffalo (*Syncerus caffer*) and Western Cattle Egrets (*Bubulcus ibis*)

Capped Wheatear (*Oenanthe pileata*)

ADDO SELF-DRIVE

Routes, Roads & Ratings

Text by
Ingrid van den Berg

Photography by
Philip & Ingrid van den Berg
Jean du Plessis

Additional photography by
Heinrich van den Berg, Gerhard Vosloo
& Karin Mitten

Published by
HPH Publishing

Spotted Hyena (*Crocuta crocuta*)

Bokmakierie (*Telophorus zeylonus*)

Contents

African Savanna Elephant *(Loxodonta africana)*

Foreword

We don't inherent the Earth from our ancestors, we borrow it from our children.
– Native American proverb.

Short-sightedness is one of conservation's biggest obstacles. When you forget about the real reason you are doing conservation – whether it is protecting animals or plants or conducting research in order to save it from extinction – the goal must always be to leave it in a better space or to make sure it can keep functioning as it was meant to be. And to do that you need to start from the bottom up.

The same rules apply when visiting Addo Elephant National Park. Start at ground zero and then work your way up. You will not only understand the bigger picture better, but you will appreciate the smaller things more when you realise the giants are actually walking on the shoulders of dung beetles and tortoises. That the big ocean dwellers like whales and sharks value the eyesight of marine birds more than what us as humans value friendship, and they have long ago learnt success comes from working together.

The idea behind the proclamation of Addo wasn't only to protect people from elephants and vice versa, but also to give people the chance to see the different ecosystems and how they naturally blend into each other: from Albany thicket to Succulent Karoo to Marine. All one park but uniquely in their different composition. Addo is also in the privileged position that it is still expanding through the purchases of land adjacent to the park and to try and consolidate the parks boundary. For people to participate in conservation they must understand and appreciate the delicate network that links these different biomes. Only then will they understand why SANParks' staff have to collar animals such as lions and elephants, translocate animals or have elephant exclosures at certain waterholes.

Tourism isn't just people driving around but rather a case of teaching of people what conservation means and you can only do that by making information easily accessible and understandable. SANParks tries to do just that through publications like this. It is only when you understand something that you can meaningfully participate in it.

Parks have the unasked responsibility to teach people that in order for us all – people and animals – to co-exist, we need to find common ground or even just a common path to ensure that all giants can keep walking or swimming on this ever-shrinking Earth.

Nick de Goede
Park Manager: Addo Elephant National Park

Spotted Hyena (*Crocuta crocuta*)

Introduction

African Savanna Elephant *(Loxodonta africana)*

Introduction

For centuries, wilderness areas were not highly valued, until people started realising such areas were disappearing at an alarming rate. This was also the case with the Addo area, but a host of factors seemed to influence its development. Perhaps owing to the rugged terrain and impenetrable bush that deterred extensive road construction and farming, together with the foresight of remarkable role-players such as the South African National Parks (SANParks), private donor agencies, local landowners and communities, big corporates, conservation organisations and the private sector, the area now boasts some of the **last remaining truly wild places** in southern Africa. The **Addo Elephant National Park** is indeed a special place with impressive landscapes and a unique set of biomes that support an incredible diversity of species, of which the African elephant is the most endearing.

Although the park was originally established to save the 11 remaining Addo elephants after the 1919–1920 massacre, the realisation slowly dawned that this area was also home to a **unique diversity of flora and fauna** and that it was imperative to conserve these exceptional ecosystems for future generations. Nowhere else in South Africa, and probably in the entire world, are so many and unique biomes represented in a single national park.

Anybody who has visited the Eastern Cape between the Kei River mouth and the town of Riversdale in the Western Cape, must have noticed the unique kind of vegetation that covers vast areas between the coastal and inland regions. This kind of vegetation used to be classified as part of the Savanna biome, but is now acknowledged as **Subtropical Thicket**, a separate biome in its own right. Pristine thicket is impenetrable woody vegetation varying from low shrubland to low forest and imbues the area with a permanent blue-green hue. Twenty percent of the recorded 1 558 plant species is endemic and found nowhere else in the world. Many associate this kind of vegetation with a huge variety of succulents and vicious thorns. Little wonder the early hunters described it as 'hunter's hell'. Much of the vegetation has been cleared over the years to make way for agriculture and grazing, in spite of its impenetrable nature.

The **Main Game Area** and the adjacent **Colchester Section** are conservation areas where thicket dominates. Before the establishment of the park, several areas within these sections were cleared for farming purposes and consequently reverted to **grassland**. Regrowth of the thicket is encouraged but it is estimated it may take another century or two to be restored.

The **Zuurberg** and **Kabouga sections** include some of the most inaccessible and pristine **forests** in South Africa. The Zuurberg Mountain Range forms part of the Cape Fold Belt and apart from the forests on the south-facing slopes and valleys, is covered by **grassland** and **fynbos** on the higher lying areas. Fortunately, these forests have largely been untouched by man because they were already part of forest reserves when the demand for firewood escalated as Gqeberha (Port Elizabeth) expanded.

Pelargonium *(Pelargonium peltatum)*

The **Zuurberg Forests** are not the only forests in the park. In the **Woody Cape Section** along the coast there are still huge sections of pristine indigenous forests that are quite different in composition from those in the Zuurberg Section or the well-known Knysna Forests. Where the coastline of Algoa Bay fringes the land, vast coastal dunefields have developed. These are well known among the locals and have been a conservation area for many years, but few people know that it is one of the most spectacular coastal dunefields in the world.

The **Marine Section** of Woody Cape stretches over most of Algoa Bay and includes Bird Island, St Croix and a few more small islands. The main conservation objectives are to promote the preservation of seabirds, the last viable population of abalone stock and to promote the recovery of line-fish stocks.

Beyond the mountains, the park extends into the **Darlington Section**. This part of the park comprises an area of 52 000 hectares and includes the huge Darlington Dam that supplies irrigation water to the Sundays River Valley and Gqeberha. The area falls within the **Nama Karoo biome** and the vegetation is sporadically dominated by an extraordinary Euphorbia species locally known as **noorsboom**.

> Greater Addo Elephant National Park covers a region that stretches from the fringes of the Great Karoo Basin, over the Zuurberg Mountains, across the Sundays River Valley, along hilly country covered by carpets of impenetrable blue-green thicket, all the way to the coast where the Sundays River forms an impressive estuary before it spills into Algoa Bay. From the mouth eastwards, a spectacular dune field forms a sand barrier between the coastline and the unique coastal forests of Alexandria. The Marine Section, which includes the Bird Island and St Croix Island groups, has recently been added.

Location map of Addo Elephant National Park

The primary function of all national parks is to **conserve biodiversity**. On the other hand, parks are also **there for people** and consequently have to make provision for visitor amenities, ensure adjacent communities derive benefits from the park and allow local economies to flourish because of the park. SANParks has adopted a **multiple zoning system** for all its parks. Wilderness zones, where human footprints are kept to the minimum, are accompanied by areas where visitors can enjoy wilderness qualities, game-drive zones where visitors can marvel at the species diversity and natural landscapes, zones for nature-based activities, zones where people can socialise, sleep over and relax, picnic areas for day visitors, etc.

PLEASE NOTE

- Refrain from littering. Keep trash in your vehicle until you can safely dispose of it in a trash can back at camp. Plastic kills. Keep it out of the park.
- Do not feed animals.
- The speed limit is 40 km/h. Maximise your game viewing and keep to between 20 and 25 km/h.
- Animals have the right of way. Never harass them.

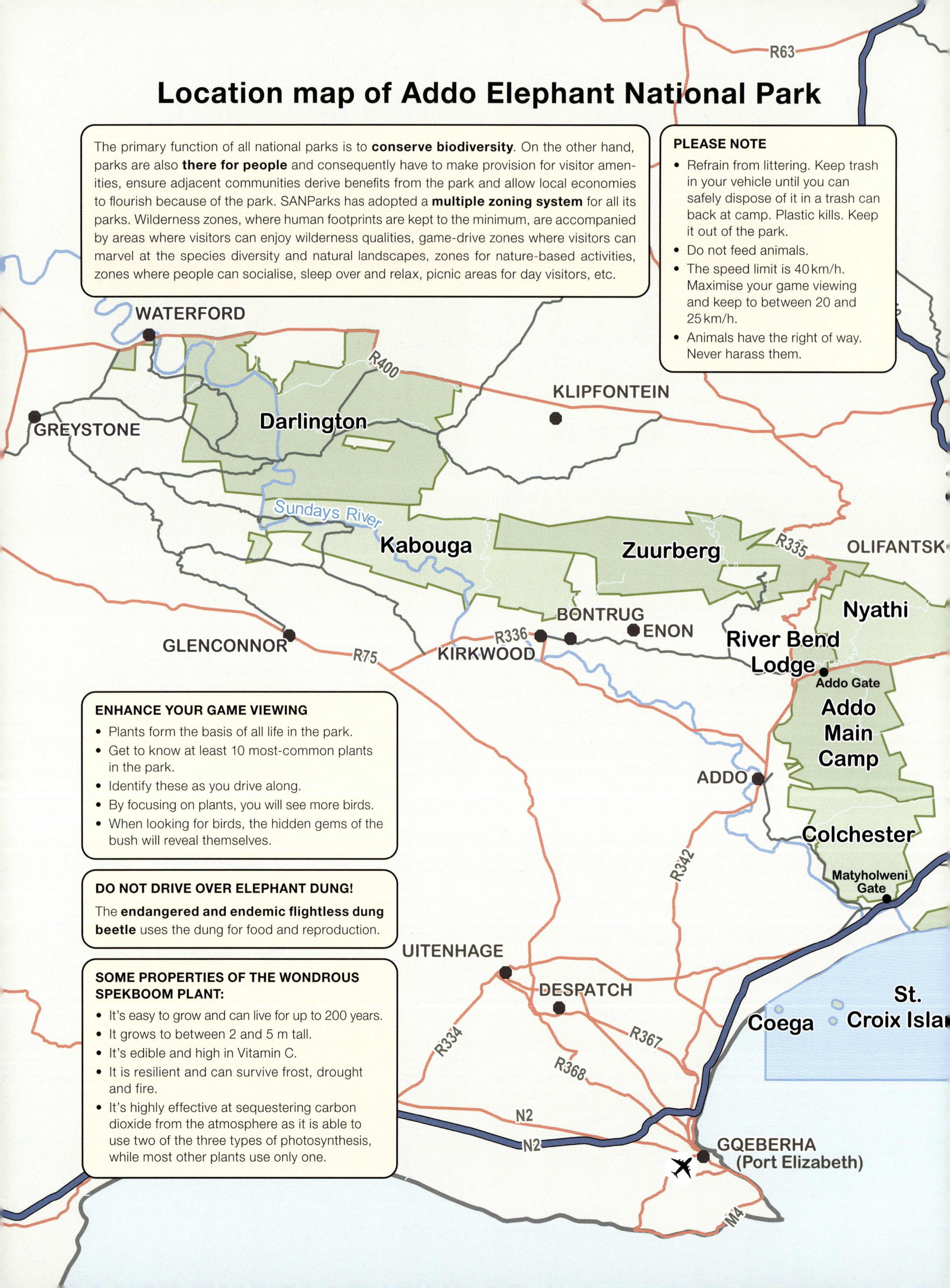

ENHANCE YOUR GAME VIEWING

- Plants form the basis of all life in the park.
- Get to know at least 10 most-common plants in the park.
- Identify these as you drive along.
- By focusing on plants, you will see more birds.
- When looking for birds, the hidden gems of the bush will reveal themselves.

DO NOT DRIVE OVER ELEPHANT DUNG!

The **endangered and endemic flightless dung beetle** uses the dung for food and reproduction.

SOME PROPERTIES OF THE WONDROUS SPEKBOOM PLANT:

- It's easy to grow and can live for up to 200 years.
- It grows to between 2 and 5 m tall.
- It's edible and high in Vitamin C.
- It is resilient and can survive frost, drought and fire.
- It's highly effective at sequestering carbon dioxide from the atmosphere as it is able to use two of the three types of photosynthesis, while most other plants use only one.

PARK ENTRANCE

Gate times: 07:00 to 18:00

GAME AREA ENTRANCE

Gate times for residents: Changes according to season. No driving after stipulated closing times.

• October–March	Summer	05:30–18:30
• April–May	Autumn	06:00–18:00
• June–July	Winter	06:30–18:00
• August–September	Spring	06:00–18:00

DESIGNATED LOOKOUT POINTS

Leaving your vehicle or hanging out of windows or sky-light openings are not allowed, unless it is at a designated lookout point. Be careful and look around when you leave your vehicle (at own risk). There are dangerous animals in the park. Lookout points are **Domkrag**; **Zuurkop**; **Kadouw**; **Spekboom**; **Algoa Bay** and **Ndlovu**.

You can safely get out of your vehicle at **Jack's Picnic Site** in the Botanical Reserve where there are picnic tables and bunks as well as toilet facilities.

QUICK FACTS

Addo Elephant National Park:

- is the third largest national park in South Africa;
- covers a total area of 178 918 ha (includes the marine section);
- is still developing, new areas will be added in due course;
- is renowned for its excellent elephant viewing;
- includes spectacular land- and seascapes;
- has both terrestrial and marine components;
- has an amazingly diverse fauna and flora;
- includes five of the seven terrestrial biomes in South Africa;
- includes 88 km of marine transgressive dunefields largest in the southern hemisphere;
- is rich in cultural heritage;
- is malaria-free; and
- ranges in altitude from 0 to 951 m.

MORE THAN JUST ELEPHANTS

Your main reason for visiting may have been to see the elephants... **BUT** Addo has much more to offer!

It is not only home to the BIG FIVE, but also to the **BIG SEVEN**!

- **African elephant**, the biggest land mammal on Earth
- **Lion**, icon of Africa
- **Leopard**, the most elusive of them all
- **Cape buffalo**, the most feared by hunters
- **Black rhino**, the highly endangered
- **Whale**, the largest living mammal
- **Great white shark**, the ocean's most adept predatory fish

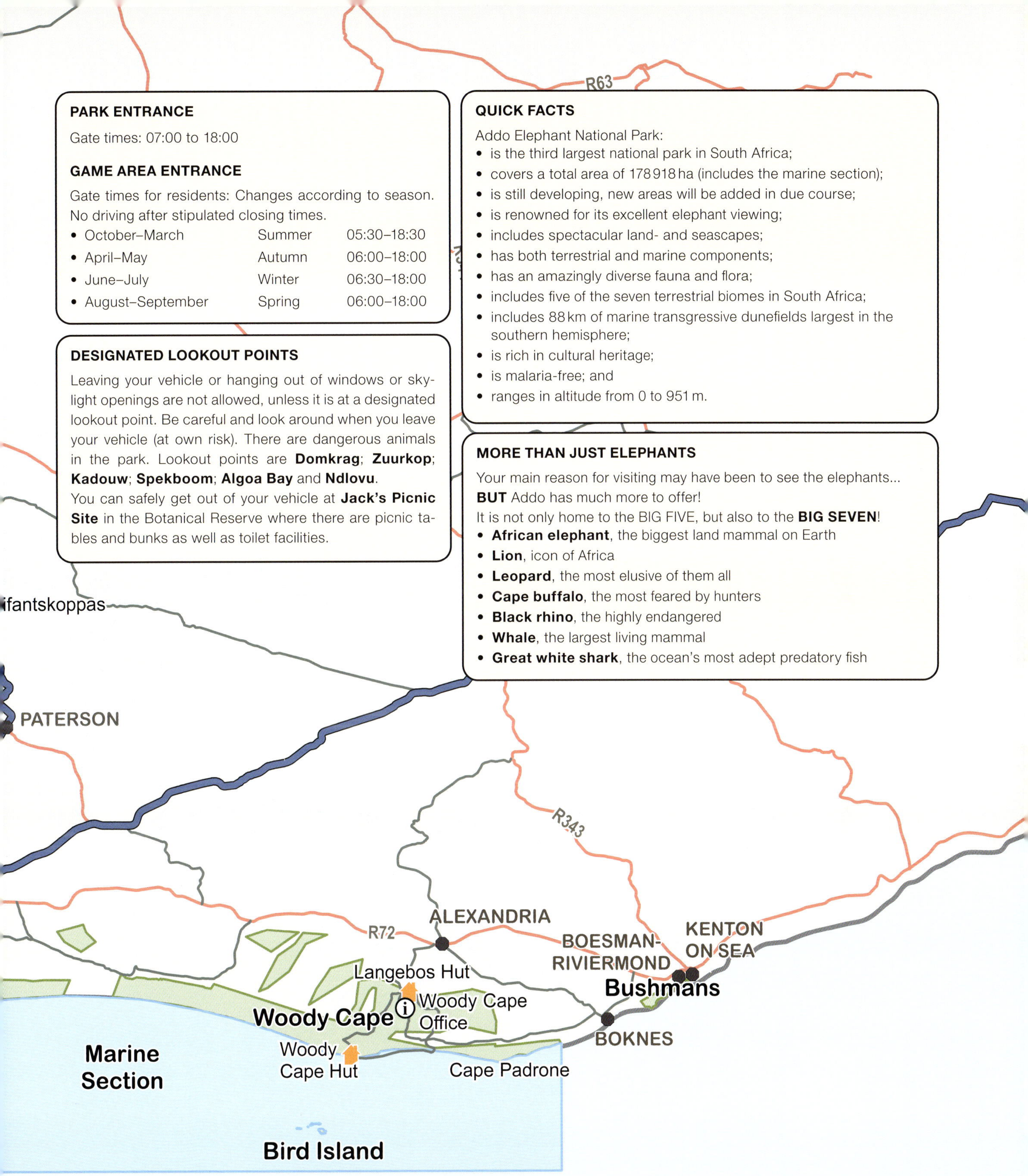

African Savanna Elephant *(Loxodonta africana)* dust bathing

Directions to the park

By air

The nearest **international airport** is in Gqeberha (Port Elizabeth), 75 km from the park. International airports are situated at Johannesburg and Cape Town, with connecting flights to Gqeberha available. **Cars can be hired** from various agencies at the airport in Gqeberha.

Visit the Nelson Mandela Bay Tourism website for more information: *www.nmbt.co.za*

By coach tour

Several **coach operators** have organised tours to the park. Visit the Nelson Mandela Bay Tourism website: *www.nmbt.co.za.*

Directions from Gqeberha (Port Elizabeth)

Shortest route from Gqeberha (Not recommended)

Take the N2 highway towards Makhanda (Grahamstown). Take the turn-off to the left signposted 'Motherwell' and 'Addo Elephant National Park'. Turn left at the top of the off-ramp and proceed through Motherwell. Follow this road until you see the entrance to the park on the right. **Distance from Gqeberha to the park – about 72 km**. This is the shortest route.

Alternative route from Gqeberha

From Gqeberha via R342 (recommended)

Take the N2 highway towards Makhanda (Grahamstown) and carry straight on until the N2 splits off to the left. Continue straight, following the N10 towards Cradock/Cookhouse. Take the R342 to the left when you get to the intersection with Paterson on your right. There is a signpost, 'Addo Elephant National Park'. Follow this road, looking out for the entrance to the park on your left. **Distance from Gqeberha to the park – about 92 km**. This is the best route, especially if you are towing a trailer or caravan.

Second alternative route from Gqeberha

From Gqeberha via Matyholweni Gate

Take the N2 highway towards Makhanda (Grahamstown) and continue straight until you pass the bridge over the Sundays River and Colchester looms ahead. Turn off to the left at the sign 'Addo Elephant National Park' and follow the signpost to the Matyholweni Gate at GPS: S33°40'57" E25°47'44". Enter the park and follow the road to Main Camp. **Distance from Gqeberha to Main Camp – about 73 km**. No caravans, trailers or buses allowed to enter here.

Directions from Makhanda (Grahamstown)

Take the N2 towards Gqeberha. Follow this road until you see the turn-off to the N10 towards Cradock/Cookhouse. Take this turn-off and continue straight, following the N10 towards Cradock/Cookhouse. Take the R342 to the left when you get to the intersection with Paterson on your right. There will be a sign posted 'Addo Elephant National Park'. Follow this road, looking out for the entrance to the park on your left.

Directions from Cradock

Follow the N10 towards Paterson. Take the R342 to the right when you get to the intersection with Paterson on your left. There will be sign posted 'Addo Elephant National Park'. Follow this road, looking out for the entrance to the park on your left.

Main Gate: S33°26'30" E25°45'2"

SANPARKS DOES NOT ACCEPT CASH.
Speedpoints are available at all sections and gates.

Ostrich chased by playful Elephant calf.

Warthog (*Phacochoerus africanus*)

Terrestrial and marine sections

The terrestrial zone

The climatic zones and geologically varying landscapes of South Africa give rise to distinct kinds of vegetation, which in turn form the basis of animal diversity and distribution. Taking all this into account, **eight different biomes** can be distinguished. They are: Fynbos, Forest, Grassland, Savanna, Subtropical Thicket, Succulent Karoo, Nama Karoo and Desert.

A **biome** is a region with a similar geology, where climate and soil types are much the same. This brings about a grouping, or a community, of plant types that can tolerate these specific climatic conditions and thrive within the given soil types. Plants support animals, either directly or indirectly. Some plants and animals occur only in specific biomes, while others may occur in more than one biome because they can tolerate a wide range of environmental conditions.

Five biomes are conserved in the terrestrial section of **Addo Elephant National Park**.

- The largest biome is **Subtropical Thicket** that covers 69% of the conservation area and occurs in the original Addo Section, the Colchester Section, Kabouga and Nyathi sections.
- **Forest** occurs in the Woody Cape Section (Alexandria Forest) and Kabouga Section (Afromontane forests) and covers 10% of the park.
- **Nama Karoo** encompasses 7% of the total area and is incorporated in the Darlington Section.
- **Fynbos** occurs on the mountain plateaus of the Zuurberg Range and covers 5% of the park.
- **Grassland** covers 5% of the park, occurs on mountain slopes and plateaus of Zuurberg and Kabouga, as well as on disturbed patches in the Main Game Area and Colchester where rehabilitation is encouraged.

Alexandria Dunefield and Marine Section

The Marine Protected Area (MPA)

The Marine Protected Area forms part of the Addo Elephant National Park (AENP) and basically encompasses the Bird Island group (Bird, Stag, Seal and Black Rocks Islands) as well as the St Croix Island group (St Croix, Jahleel and Brenton Islands). The MPA has now been enlarged to cover 120 000 ha from Canon Rocks in the east to Coega in the west.

User zones

User zones serve to ensure that high-intensity activities and facilities occur in areas that will tolerate intensive use. In this way sensitive areas are protected from over-utilisation.

Huge sections of the AENP are presently zoned as **wilderness areas**. The main aim in these areas is to conserve the natural environment without ecological disturbances and human access is limited. Mountains and hills make up most of these wilderness areas. As the park expands, more farmland is incorporated and rehabilitation of disturbed areas commences. Disturbed areas take centuries to rehabilitate.

The sections designated for **general tourist activities** have an extensive road network. The roads were planned with great care to avoid disturbing more of the existing natural vegetation and to comply with ecological principles. Roads in the park are suitable for all vehicles but special permission from the park manager is required for open-top game-viewer vehicles.

The main access roads into the park are tarred, or at least partly tarred. Sections of gravel roads ascending steep hills are also tarred to prevent erosion. The gravel roads are usually in a good condition but when the weather is wet, gravel roads may be closed. The Southern Access Road links Matyholweni Gate and Camp to Addo Main Camp (36 km) but is not accessible to buses and cars pulling caravans or trailers, unless they are designed for off-road purposes. Red or greenish-grey mudstone of the Sundays River Formation underlies most of the game park and this type of soil weathers to form clay, which is extremely slippery when wet.

Tourist user zones for self-drives

The Addo Elephant National Park or AENP is a **park in progress** and although it encompasses such an immense area, much of it is not accessible to self-drive tourism. The following areas are suitable for self-drive visitors:

- The **Main Game Area** accessible from Main Camp and from the Colchester Section.
- The **Colchester Section**, which is an extension of the Main Game Area and is also accessible from the Matyholweni Entrance Gate in the south.
- The **access road to the Nyathi Rest Camp** (for residents only) which traverses the River Bend concession area.
- The **public road R335** (not part of the park) leading to the Zuurberg Section for accessing hiking trails on the mountain and the secluded Narina Tented Camp.
- The **Kabouga Section** for scenic- and game viewing and access to the 4x4 Bedrogfontein route.
- The **Darlington Section** for limited game viewing along the public road R400.
- **Public road from Alexandria to the Woody Cape Office** to access the hiking trails through the forest.

Marion Baree Waterhole and African Savanna Elephant (*Loxodonta africana*)

Domkrag Waterhole

Lion *(Panthera leo)* at Marion Baree Waterhole

Routes

Routes

in the Main Game Area and Colchester Section

Choosing a route

Game-viewing areas and the road network

The primary self-drive **game-viewing areas** in the Addo Elephant National Park are the **Main Game Area** (about 13 000 ha) and the **Colchester Section** (13 000 ha). Here you can expect to see the famous Addo elephants, some of Africa's most impressive predators such as lion and hyena, other giants such as the endangered black rhino and stately buffalo, zebra, warthog, primates and a variety of antelope and birds.

Addo is, however, not only about elephants or game animals, but also about a very special plant community or biome called **Subtropical Thicket** (this vegetation type is sometimes also called Albany Thicket or Addo Bush). This unique vegetation belt occurs in parts of the Eastern Cape and is neither savanna nor forest, although it intermittently resembles both. Thicket interspersed with grassland patches is the dominant vegetation type in the game-viewing area.

The Main Game Area opened to the public in 1978 while the fence to the Colchester Section was dropped only in 2006. These two sections form a perfectly functional ecological unit resulting from the purchase and rehabilitation of several farms south of Addo Heights.

The road network may sometimes be overwhelming for first-time visitors and even for those who haven't visited in some time. The section on the routes will help you decide and plan, while the section on the roads will give more information on what you may expect to find.

All visitors to the park enter the game-drive area through one of the two entrance gates and/or camps. Two sets of routes are suggested, starting at each of the entrance gates. Study the map and recommended routes provided but **maintain flexibility**. You may want to chop and change. Most routes start and end at the same gate but there are exceptions.

Some of the roads have established names but others can be identified only by checking the numbers at the intersections. One example is the Link Road between the Gorah Loop and the Zuurkop Road. Check the map and the table to establish where the different roads are and how they are referred to in this book.

The roads are listed below in **alphabetical order**. Find the roads on the map. Each intersection in the park is numbered from one to 40. These **points** are useful in finding your way.

Road name	Points	Rating	Page	Condition	Distance
Access Road from Main Camp Gate to first point	1	★★★	p.50	Tarred	1.5 km
Access Road from Matyholweni Gate to Addo Heights	36, 35, 34, 33, 32, 31,30, 29, 40, 38, 28	★★★★	p.52	Gravel and tar	18.2 km
Domkrag and Nzipondo Loop	1, 2, Nzipondo 2, 3	★★★★★	p.58	Tarred	1.9 km
Gorah Loop	3, 8, 9, 7, 8, 3	★★★★	p.62	Gravel	12.7 km
Gwarrie and Rooidam	11, 12, 13	★★★★★	p.70	Tarred	2 km
Hapoor Dam via gravel road	13, 22	★★★★	p.76	Gravel and tar	7.5 km
Hapoor Dam via Main Tarred Road	5, 11, 14, 21, 22	★★★	p.80	Tarred	11.5 km
Harvey's Loop	37, 29	★★★★	p.86	Gravel	15.9 km
Jack's Picnic Site and Botanical Reserve	25	★★★★	p.90	Gravel	1 km
Link Road	7, 6	★★★★★	p.92	Gravel and tar	1.3 km
Mbabala Loop	16, 17, 18, 18	★★★	p.94	Gravel	7.3 km
Mbotyi Loop	34, 32	★★★★	p.96	Gravel	4.6 km
Mpunzi Loop	22, 23	★★★	p.98	Gravel	2.6 km
Ngulube Loop	38, 31	★★★★★	p.100	Gravel	10.4 km
Spekboom Road	21, 23, 24, 25, 26, 27	★★★★	p.104	Gravel and tar	9.6 km
Vukani Loop	30, 33	★★★★	p.110	Gravel	4 km
Woodlands Road	3, 4, 5, 1	★★★★★	p.112	Tarred	3.1 km
Zuurkop Road	4, 6, 10, 19, 20, 15, 14	★★★	p.116	Gravel and tar	7.4 km

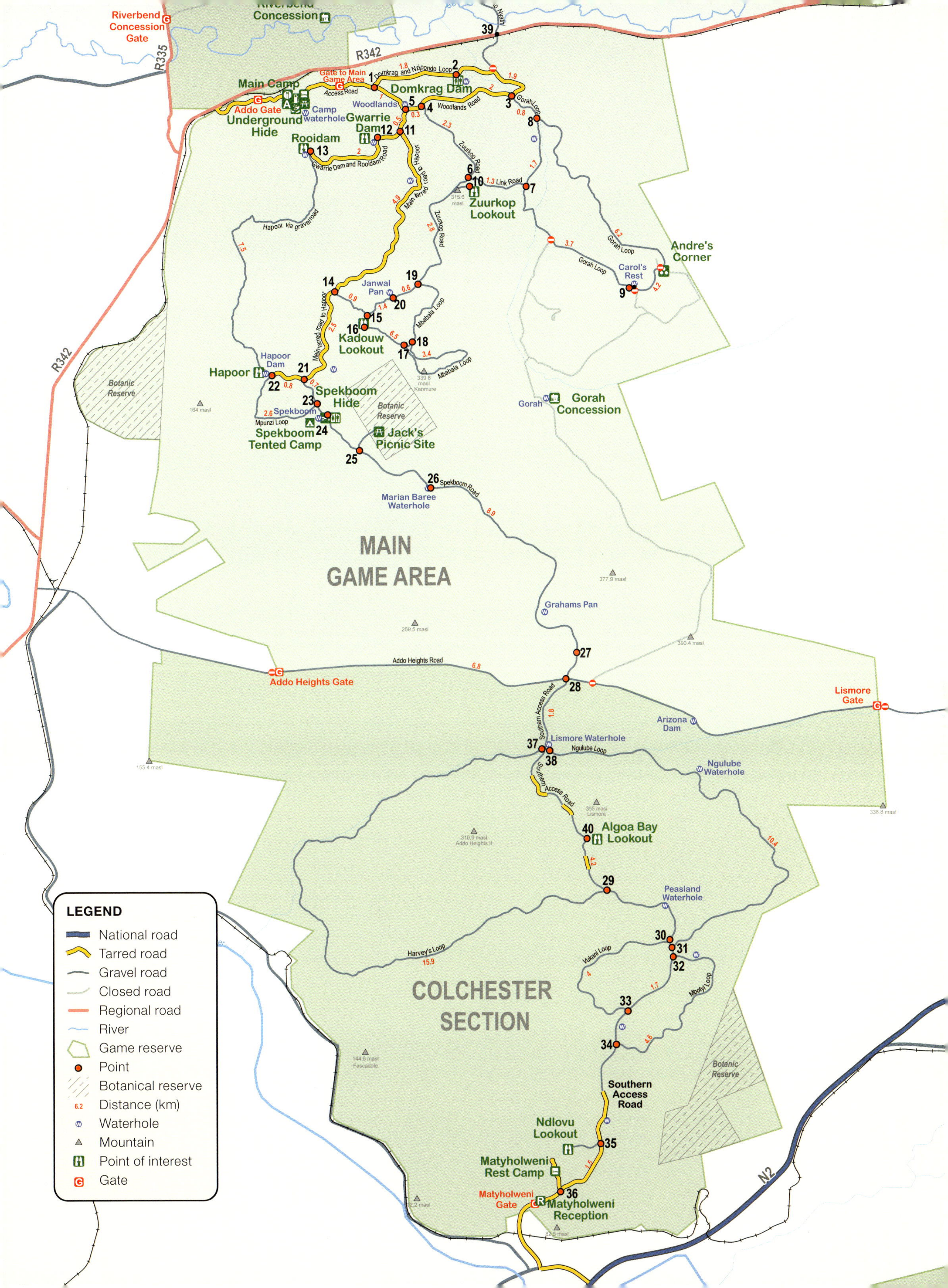

Riverbend Concession Gate
Riverbend Concession
R335
R342
Main Camp
Addo Gate
Underground Hide
Camp waterhole
Access Road
Gate to Main Game Area
Domkrag and Nzipondo Loop
Domkrag Dam
Woodlands
Woodlands Road
Gwarrie Dam
Rooidam
Gwarrie Dam and Rooidam Road
Gorah Loop
Zuurkop Road
Link Road
Zuurkop Lookout
Main tarred road to Hapoor
Hapoor via gravel road
Andre's Corner
Carol's Rest
Janwal Pan
Mbabala Loop
Kadouw Lookout
Hapoor Dam
Hapoor
Botanic Reserve
Spekboom Hide
Spekboom
Mpunzi Loop
Spekboom Tented Camp
Jack's Picnic Site
Gorah
Gorah Concession
Marian Baree Waterhole
Spekboom Road
MAIN GAME AREA
Grahams Pan
Addo Heights Road
Addo Heights Gate
Lismore Gate
Arizona Dam
Southern Access Road
Lismore Waterhole
Ngulube Loop
Ngulube Waterhole
Algoa Bay Lookout
Peasland Waterhole
Harvey's Loop
Vukani Loop
Mbotyi Loop
COLCHESTER SECTION
Southern Access Road
Ndlovu Lookout
Matyholweni Rest Camp
Matyholweni Gate
Matyholweni Reception
N2
LEGEND
National road
Tarred road
Gravel road
Closed road
Regional road
River
Game reserve
Point
Botanical reserve
Distance (km)
Waterhole
Mountain
Point of interest
Gate

Ratings and what they are based on

Ratings of the different roads and sections of roads **are subjective** and meant to be mere indications of what to expect. They are not cast in stone, nor are they based on any scientific findings. Several important **factors influence and determine wildlife sightings** in the Addo Elephant National Park (AENP). Sightings experienced on a specific road the one day may completely differ from those experienced the very next day on the same road. The ratings do not necessarily pertain to elephant sightings because elephants seem to be everywhere and nowhere simultaneously. Now you see them, the next moment you don't. Sightings of elephants and most other animals are totally unpredictable. The presence of some game may be predictable up to a certain extent, perhaps because they are seen regularly along a certain road, or because there is a den close by.

The ratings are based on the following:

- Thicket tends to be monotonous and there is little variation, but some areas are more **scenic** than others.
- In some places the thicket is denser and in prime condition, therefore **condition of the thicket vegetation** is another consideration.
- The park has several open grassland areas (formerly cleared mainly for farming purposes) or areas with karroid vegetation, which attract different species. Elephants and buffalo can show up on any road, in any area but there are areas where the likelihood of breeding herd sightings is higher. **Species diversity** is thus taken into account. This pertains mainly to mammals and birds.
- **Visibility** is another factor. Thicket is dense and animals disappear in a flash.

African Savanna Elephant *(Loxodonta africana)*

- **Road condition** plays a role; tarred roads versus gravel roads and tracks.

Therefore, **where a road is rated as 5 star** (★★★★★) it means it is **scenic**, with thicket in **prime condition**, offers a **high species diversity**, **good visibility** and the road is **tarred**.

How to plan your route

- Study the area map.
- Decide between a morning drive (short or long) and/or afternoon drive (usually short) or a day drive (usually day visitors).
- Look at breakfast and/or lunch options or pack a picnic basket.
- Look at the recommended routes, read more about them in the chapter on roads to make informed decisions.
- Locate the hotspots and animal drinking places on the area map.
- Decide on your route by taking all the above into account.
- Four to five hours for a morning drive is more than enough. An afternoon drive would usually not exceed two to three hours. A day drive should not exceed 150 km. Allow time for watching elephants and other game at a waterhole or relaxing at a picnic site. Covering too great a distance will leave you exhausted and inclined to break the speed limit.
- The suggested routes show only the main points, with a rough calculation of the time it would take for an unhurried game drive. Calculate approximate time, allowing for picnic breaks and time to spend at excellent sightings.

Weather conditions and how they can influence game sightings

Weather varies from day to day and from one season to the other.

- Day **temperatures** definitely play a huge role – when it's hot, more animals will congregate at the waterholes and vice versa. Elephants and most other game will seek shade during the heat of the day. All the mega herbivores will probably visit a waterhole not only for drinking, but also for wallowing and dust bathing.
- On extremely **windy** days, the elephants simply vanish into the bush and none can be spotted.
- Predators are mostly **nocturnal** but may show themselves early in the morning or late in the afternoon. The park is huge, and it is possible that days and even weeks go by without predator sightings.
- **Rainfall** is important. During a drought, animals may forage differently from the norm. Exclusive grazers may revert to browsing to survive adverse conditions and elephants, which usually consume a lot of grass, spend more time browsing. After good rains when grasses sprout, elephants visit the open grasslands again and browsers are also seen munching on fresh herbs, forbs and even grass sprouts.
- **Seasons** play a role in birdsong and rutting. In spring, birdsong may be magnificent, but could disappoint during other seasons. During the rutting season antelope display behaviours normally not witnessed.
- Cloudy or sunny weather may be good or bad.
- Sunshine or rain, hot or cold, day or night – what is important is **the expectation of the unexpected, which adds to the thrill**.

Hotspots

Valleys are usually rich in minerals and moister than the hills, with Valley Thicket that includes **sweet-thorn acacia,** which is a favourite for many species. In winter, it is warmer in the valleys and they provide shelter against the wind.

The main reason why **grasslands** are regarded as hotspots is the openness and visibility they offer. Grasslands in the park are interspersed with karroid vegetation. Expect to see grazing species and grassland birds in abundance.

Since there is no natural water source in the park, water has to be pumped from underground. There are a number of **big dams and waterholes** supplying unlimited amounts of water. These are favoured by elephants and they tend to claim the dam or waterhole for themselves, leaving little space for shier species. At the same time the vegetation in the surrounding areas becomes over-utilised. It was, and still is, a huge challenge to persuade elephants to forage in another area and use other watering places, especially since elephant numbers have increased systematically over the years and elephant families have their favourite foraging and drinking places.

Presently the park is experimenting with the practice of closing off some of the drinking places for elephants by constructing a wired enclosure around some of the smaller waterholes (Carol's Rest, Lismore and Peasland). The wires are electrically charged, with flags at the height of the average elephant, but higher than other game species such as buffalo. When an elephant approaches the water, a light charge is released where some of the flags touch its face or back. This frightens most of them away, but true to elephant nature, some have already learnt to stoop when they see the string of flags!

There are several **lookout points** that offer spectacular views over the thicket-clad hills and valleys, with the extent of the blanket of Subtropical Thicket, the coast in the distance and the Zuurberg Mountain Range as backdrop.

TIPS ON SELECTING YOUR ROUTE FOR THE DAY

- Elephants visit waterholes at any time of day and night, but when it's hot they usually visit later in the morning or during the afternoon to cool down or bathe.
- Start early or stay late if you wish to see lion, hyena, jackal or caracal.
- Buffalo drink every day, either early morning or late afternoon.
- Black rhino are seldom seen and usually drink at dusk.
- The best time to see meerkat (suricates) and yellow-tailed mongooses is just after sunrise when they sun themselves at their dens before they disperse for the day.
- Kudu, bushbuck and buffalo keep to the thickets for most of the day.
- Grazers like zebra and red hartebeest are more active during the cooler times of day and keep to the grassland areas.
- Bat-eared foxes and aardwolf are nocturnal and shy and are seen only early morning or late afternoon.
- Leopard, aardvark and black-footed cat occur but are seldom seen.
- Genet and porcupine are often seen in the camps at night.
- Hares are prolific early mornings or late afternoons.
- Vervet monkeys are common and can be an irritation in the camps.
- Baboons are not common in the game-viewing areas except in hilly places, mainly in the south and in other parts of Addo.
- The Karoo bush rat is active during the day and often seen.
- Bushpig are nocturnal but warthog diurnal and common throughout the game-viewing area.
- Birding is best in the mornings or late afternoons. Grass-living birds can be seen any time of day.

Bitou Bush *(Osteospermum moniliferum)*

Greater Kudu (*Tragelaphus strepsiceros*)

Main Game Area

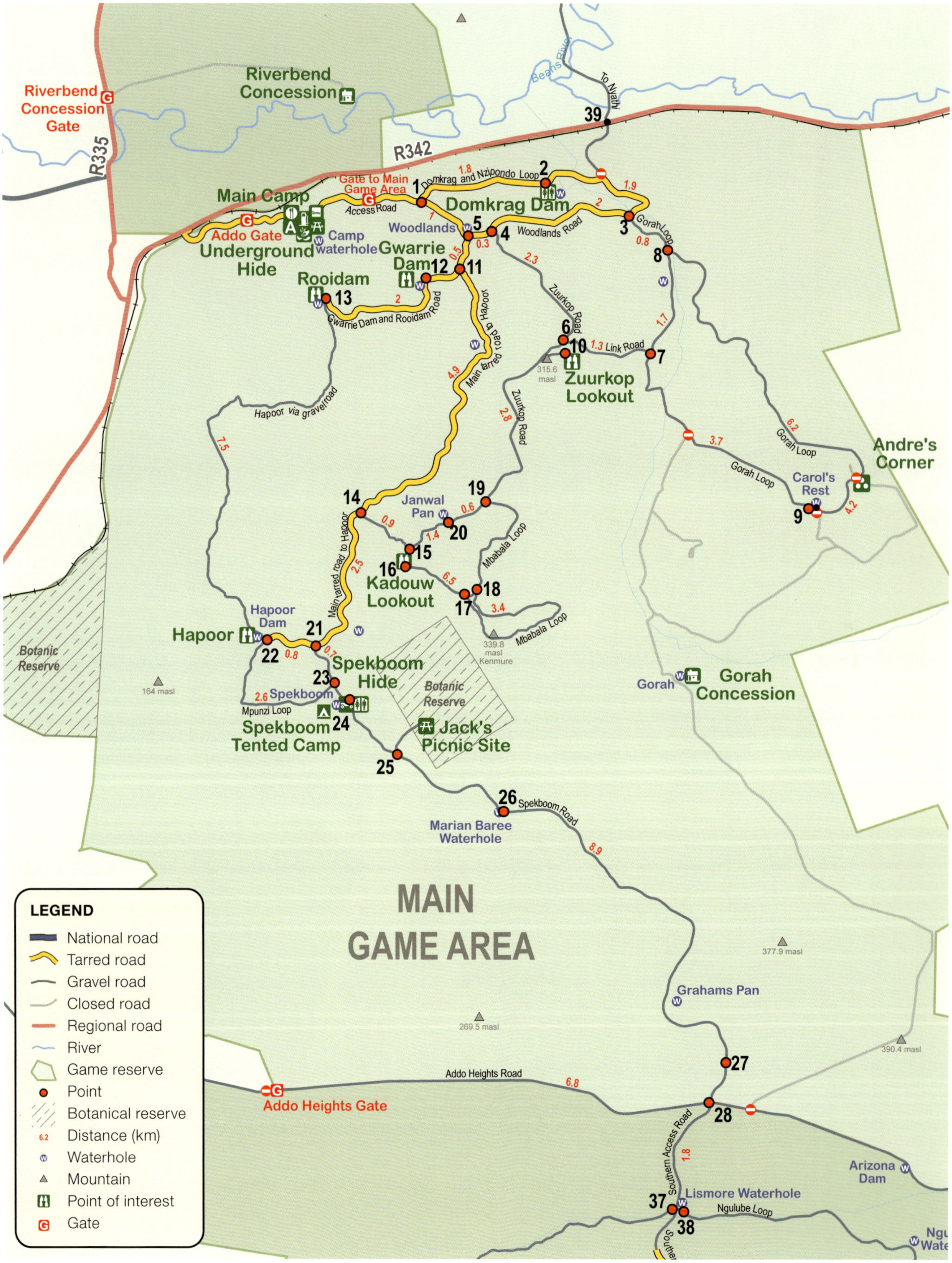

Colchester Section

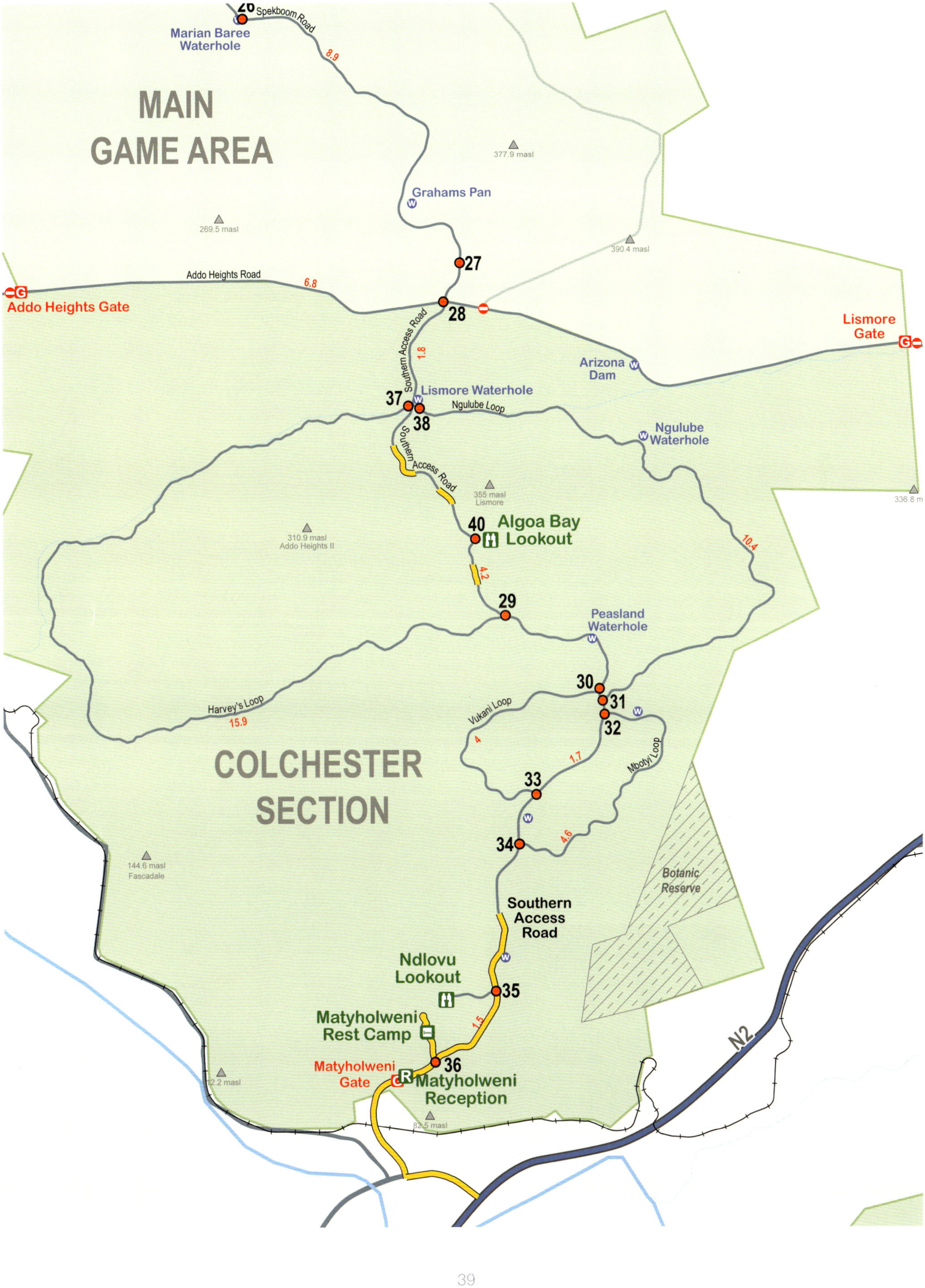

Routes starting from the Northern Access Gate

Morning drive to Gorah Loop and Hapoor

38.5 km (2–4 hours)

- Northern Access Gate
- 1, 2 to 3, Domkrag/Nzipondo Road (**Domkrag Dam**)
- 3 to 8, keep left up to 7, Gorah Loop (**Carol's Rest Waterhole**)
- 7 to 6, Link Road
- 6, 19 to 15, Zuurkop Road (**10. Zuurkop Lookout**)
- 15 to 16, **Kadouw Lookout**
- 16, 15 to 14, Main Tarred Road
- 21, 23, 24 to 25, Spekboom Road
- 25, **Jack's Picnic Site** (have a breakfast break)
- 25 to 21, back to Tarred Road
- 21 to 22, **Hapoor Dam**
- 22 to 13, Gravel Road to **Rooidam**
- 13, 12 to 11, **Rooidam** to **Gwarrie Waterhole** to Main Tarred Road
- 11 to 5, Woodlands Road
- 5, **Woodlands Waterhole** back to Main Camp

Morning drive to Addo Heights

50 km (3–5 hours)

- Northern Access Gate
- 1, 2 to 3, Domkrag/Nzipondo Road (**Domkrag Dam***)
- 3, 4 to 5, Woodlands Road (**Woodlands Waterhole***)
- 5, 11, 14 to 21, Tarred Road to the south
- 21, 23, 24, 25, 26 to 27, Spekboom Road
- (**Spekboom Hide*** and **Marion Baree Waterhole**)
- 27. Addo Heights (turn back)
- 27, 26 to 25, **Jack's Picnic Site**
- 25, 24 to 23, Mpunzi Loop
- 23 to 22, to **Hapoor Dam***
- 22 to 13, Gravel Road to Rooidam
- 13, 12 to 11, **Rooidam*** to **Gwarrie Waterhole*** to Main Tarred Road
- 11 to 5, Woodlands Road
- 5, **Woodlands Waterhole*** back to Main Camp

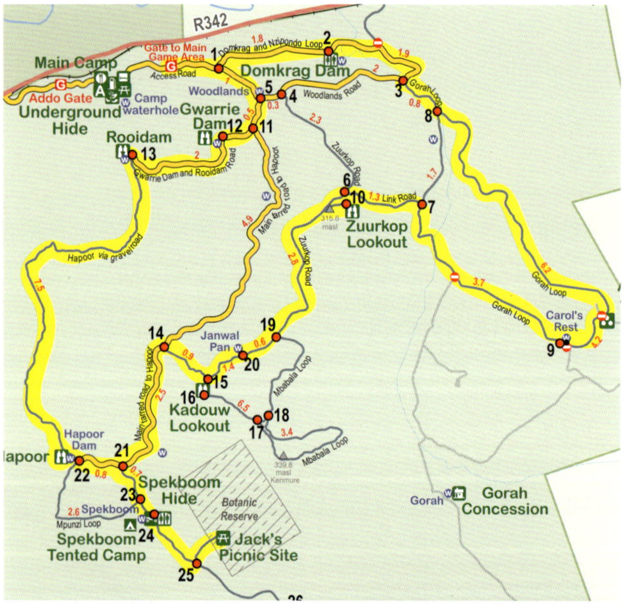

Milkbush Creeper *(Sarcostemma viminale)*

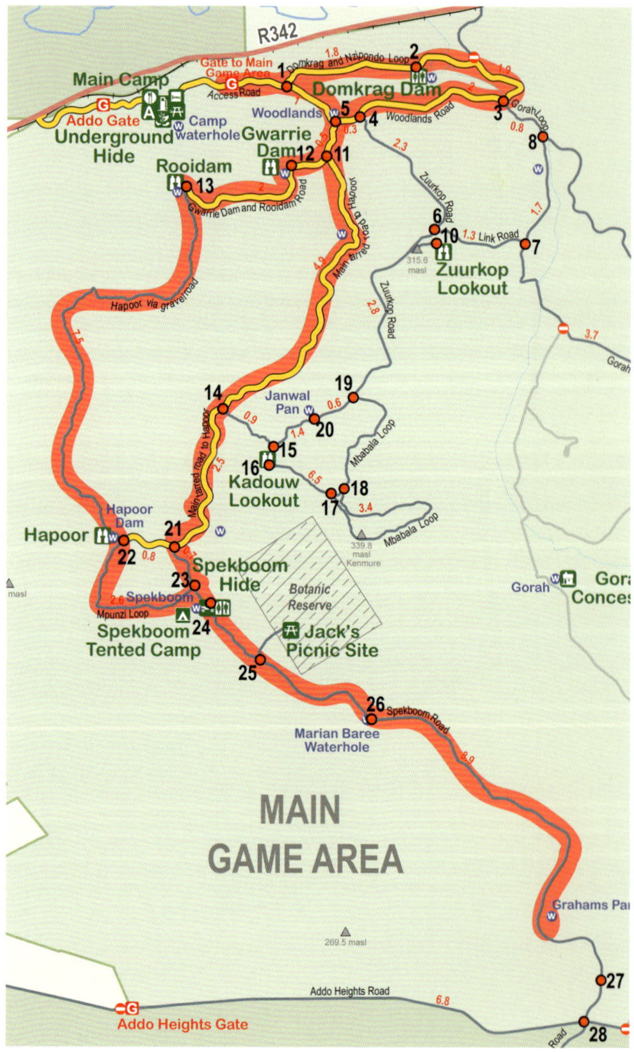

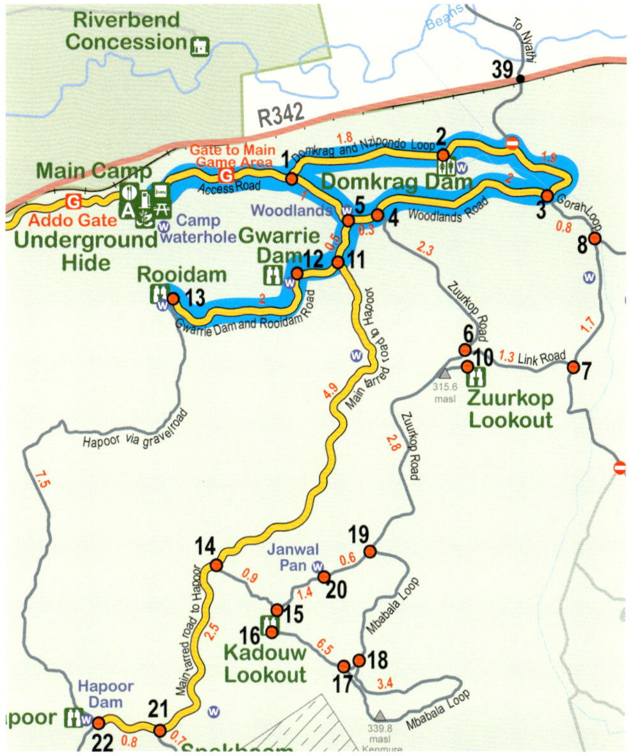

Needle Bush *(Azima tetracantha)*

African Asparagus *(Asparagus africanus)*

Route to Jack's Picnic Site

35 km (2–4 hours)

- Northern Access Gate
- 1, 5, 4 to 3, Woodlands Road (**Woodlands Waterhole***)
- 3 to 8, Gorah Loop, 8, take right split to 7, Link Road
- 7 to 6, Link Road
- 6 to 19, Zuurkop Road (**Zuurkop Lookout***)
- 19, 18, 17, 16 to 15, Mbabala Loop (**Kadouw Lookout***)
- 15 to 14, Zuurkop Road
- 14 to 21, left at Main Tarred Road
- 21, 23, 24 to 25, to **Jack's Picnic Site**
- 25 to 21, back to Main Tarred Road
- 25 to 22, left to Hapoor Dam (**Hapoor Dam***)
- 22 to 13, Gravel Road to **Rooidam**
- 13, 12 to 11, Rooidam and **Gwarrie Waterhole**
- 11 to 5, Main Tarred Road
- 5 to 1, and back to Main Camp

Very short afternoon loop

15.5 km (1–2 hours)

- Northern Access Gate
- 1, 2 to 3, Domkrag/Nzipondo Road (**Domkrag Dam***)
- 3, 4 to 5, Woodlands Road (**Woodlands Waterhole***)
- 5 to 11, Main Tarred Road to the south
- 11 to 12, **Gwarrie Waterhole***
- 12 to 13, **Rooidam***
- 13, 12 to 11, turn back towards Main Tarred Road
- 11 to 5, towards **Woodlands Waterhole**
- 5 to 1, back to Main Camp

Common Ipomoea (*Ipomoea sp.*)

Scarlet Ancestor Geranium (*Pelargonium inquinans*)

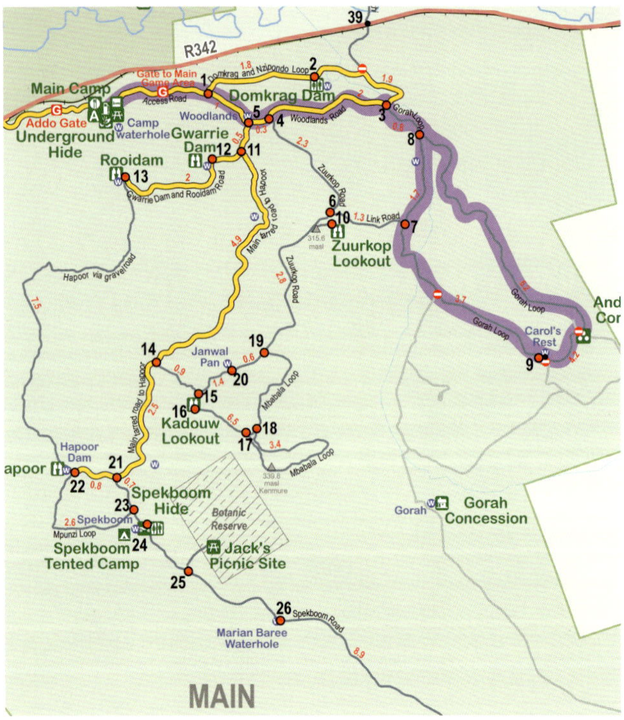

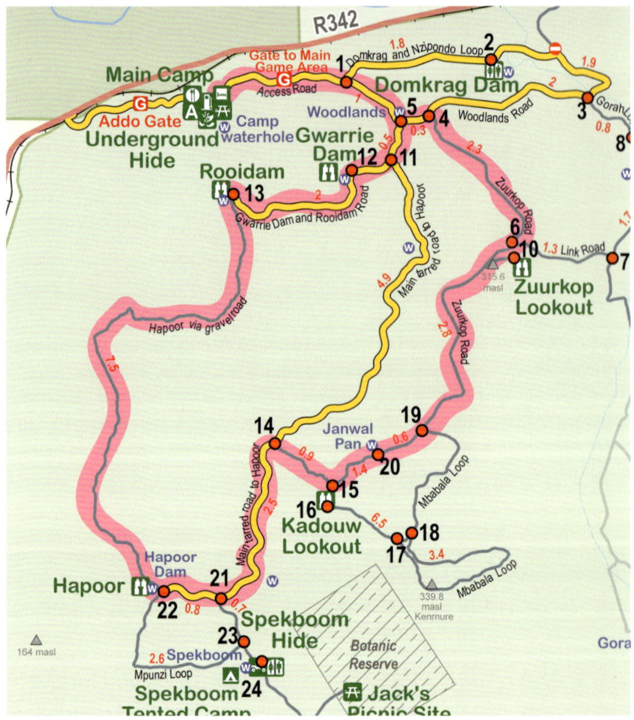

Gorah afternoon loop
16.8 km (1–2 hours)

- Northern Access Gate
- 1, 5 to 3, Woodlands Road
- 3 to 8, to Gorah Loop
- 8, 9 to 7, Gorah Loop (grasslands*, **Carol's Rest***)
- 7, 8 to 3, back to Woodlands Road
- 3, 4, 5 to 1, back to Main Camp (**Woodlands Waterhole***)

Hapoor Loop
24 km (1–2 hours)

- Northern Access Gate
- 1 to 5, **Woodlands Waterhole***
- 5 to 11, Main Tarred Road
- 11, 12 to 13, **Gwarrie Waterhole*** and Rooidam Road*
- 13 to 22, Gravel Road to Hapoor Dam (**Hapoor Dam***)
- 22 to 21, Hapoor Dam to Main Tarred Road
- 21, 14, 11 to 5, to **Woodlands Waterhole**
- 5 to 1, and back to Main Camp

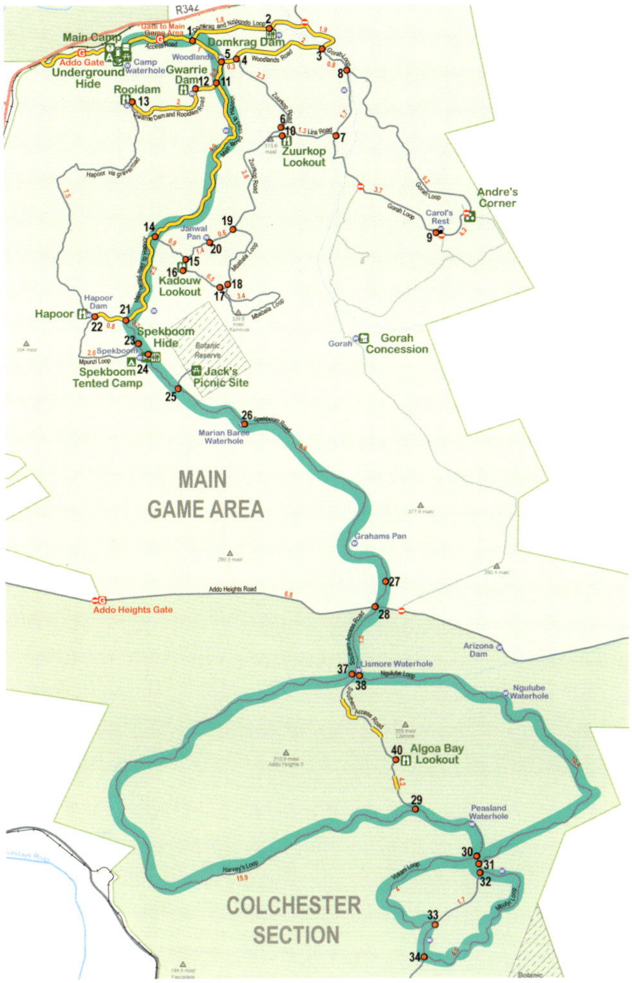

Long game drive from Main Camp to Colchester Section

78 km (3–5 hours)

- Northern Access Gate
- 1 to 5, **Woodlands Waterhole***
- 5, 11 to 21, Main Tarred Road
- 21, 23 to 24, Spekboom Road (**Spekboom Hide***)
- 24 to 25, **Jack's Picnic Site**
- 25, 26 to 27, on with Spekboom Road (**Marion Baree Waterhole***, **valley at Grahams Pan***)
- 27 to 28, pass over Addo Heights Road
- 28 to 37, Southern Access Road (**Lismore Waterhole***)
- 37 to 29, Harvey's Loop (first section*)
- 29 to 30, south on Southern Access Road (**Peaseland Waterhole***)
- 30 to 33, Vukani Loop
- 33 to 34, south on Southern Access Road
- 34 to 32, Mbotyi Loop
- 32 to 31, north on Southern Access Road
- 31 to 38, Ngulube Loop (**Ngulube Waterhole***)
- 38 to 28, Southern Access Road north
- 28, 27, 26, 25, 24, 23 to 21, Spekboom Road
- 21, 14, 11 to 5, Main Tarred Road
- 5 to 1, and back to Main Camp

Drive from Main Camp to Matyholweni from the north (or Matyholweni to Main Camp starting from the south)

39 km (2–5 hours)

- Northern Access Gate
- 1 to 5, **Woodlands Waterhole***
- 5, 11 to 21, Main Tarred Road
- 21, 23 to 24, Spekboom Road (**Spekboom Hide**)
- 24 to 25, **Jack's Picnic Site**
- 25, 26 to 27, on with Spekboom Road (**Marion Baree Waterhole***, **valley at Grahams Pan**)
- 27 to 28, pass over Addo Heights Road
- 28 to 37, Southern Access Road (**Lismore Waterhole***)
- 38 to 29, Southern Access Road (**Algoa Bay Lookout***)
- 29 to 30, Southern Access Road (**Peaseland Waterhole***)
- 30, 31, 32, 33, 34, 35 to 36, Southern Access Road
- 36, to Matyholweni Gate

Routes starting from Matyholweni

Colchester Section from the south

50 km (3–5 hours)

- Matyholweni Gate
- 36, 35, 34 to 33, Southern Access Road
- 33 to 30, turn left into Vukani Loop
- 30 to 29, north on Southern Access Road (**Peaseland Waterhole***)
- 29 to 37, turn left into Harvey's Loop
- 37, 38 to 31, Ngulube Loop (**Ngulube Waterhole***)
- 31 to 32, Southern Access Road
- 32 to 34, Mbotyi Loop
- 34 to 36, south on Southern Access Road
- Back to Matyholweni Camp or Access Gate

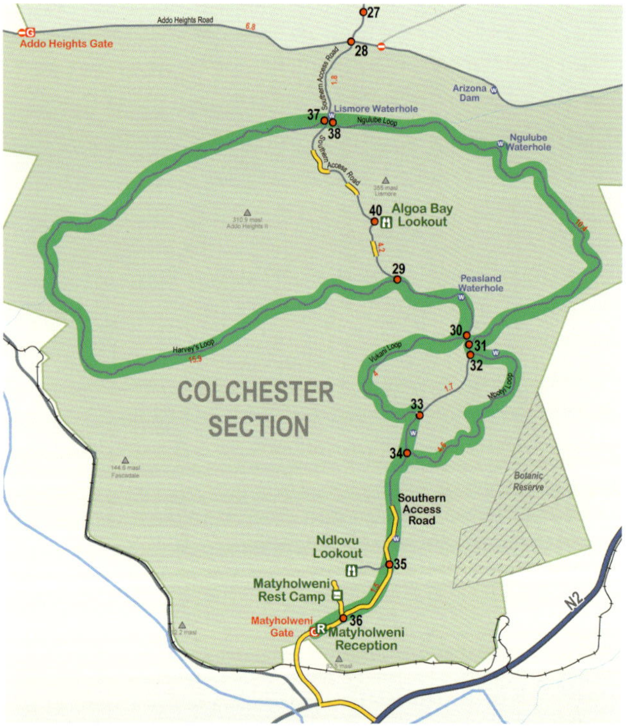

Karoo Rat

From Matyholweni Camp to Hapoor and back

59.9 km (3–5 hours)

- Matyholweni Gate
- 36, 35, 34, 33, 32, 31 to 30, Southern Access Road
- 30 to 29, north on Southern Access Road (**Peaseland Waterhole**)
- 29 to 38, north on Southern Access Road (**Algoa Lookout**)
- 38 to 28, north on Southern Access Road (**Lismore Waterhole**)
- 28 to 27, Addo Heights road crossing
- 27 to 26, north on Spekboom Road (**Grahams Pan Valley**)
- 26 to 25, north on Spekboom Road (**Marion Baree Waterhole**, **Jack's Picnic Site**)
- 25, 24, 23 to 21, north on Spekboom Road (**Spekboom Hide**)
- 21 to 22, left towards Hapoor Dam
- 22, spend time at **Hapoor Dam** and return via Spekboom Road
- 21, 23, 24, 25, 26, 27 to 28, south on Southern Access Road
- 28 to 38, Southern Access Road (**Lismore Waterhole**)
- 38 to 31, Ngulube Loop (**Ngulube Waterhole**)
- 31 to 30, north along Southern Access Road
- 30 to 33, Vukani Loop
- 33, 34, 35, 36, to Matyholweni Camp or Gate

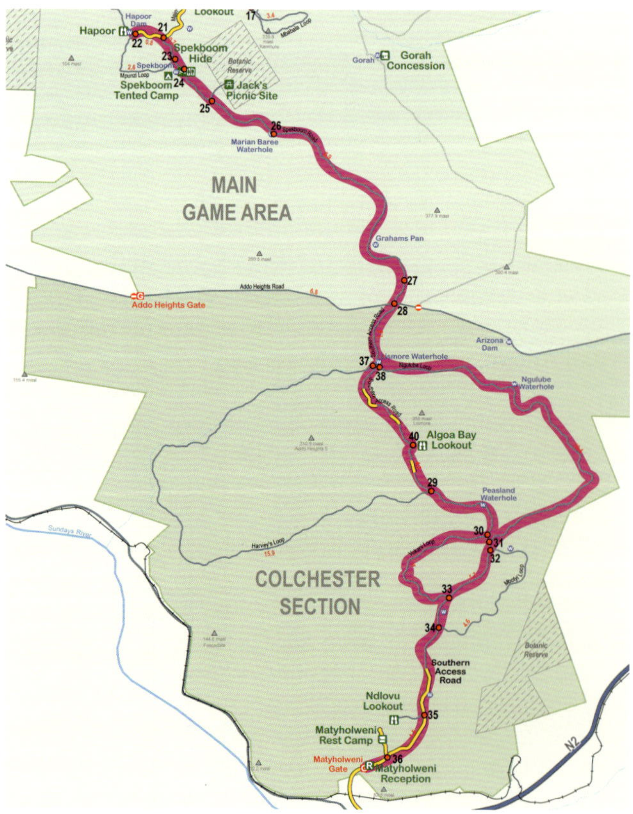

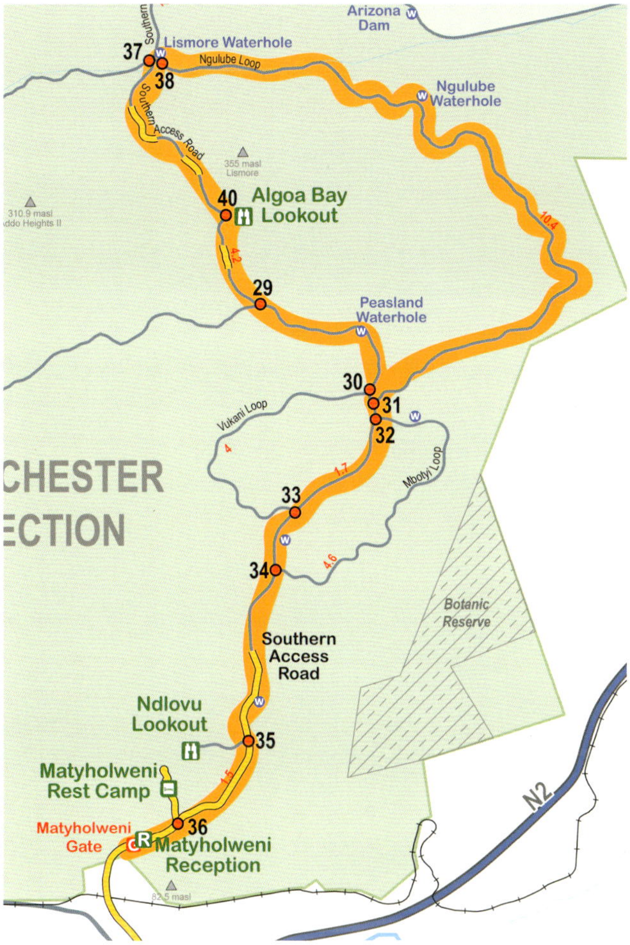

Blood/Snake Lily *(Scadoxus puniceus)*

Short afternoon drive from Matyholweni

33.4 km (2 hours)

- Matyholweni Gate
- 36, 35, 34, 33, 32, 31 to 30, north on Southern Access Road
- 30 to 29, north on Southern Access Road (**Peaseland Waterhole***)
- 29 to 38, north on Southern Access Road (***Algoa Lookout**)
- 38 to 31, Ngulube Loop (***Lismore Waterhole**, ***Ngulube Waterhole**)
- 31, 32, 33, 34, 35 to 36, south to Southern Access Road
- 36, Matyholweni Camp or Gate

Red Hartebeest (*Alcelaphus buselaphus*)

Dirt road in Addo

Roads

Roads

Note that the roads are arranged and discussed in **alphabetical order**:

- Access Road from Main Camp Gate to first point
- Access Road from Matyholweni Gate to Addo Heights
- Domkrag and Nzipondo Loop
- Gorah Loop
- Gwarrie and Rooidam
- Hapoor Dam via Gravel Road
- Harvey's Loop
- Jack's Picnic Site and Botanical Reserve
- Link Road
- Main Tarred Road to Hapoor
- Mbabala Loop
- Mbotyi Loop
- Mpunzi Loop
- Ngulube Loop
- Spekboom Road
- Vukani Loop
- Woodlands Road
- Zuurkop Road

Each road has a **star rating** along with the following details:

★★★

Road Name

Park Section

- Points at the road intersections
- Information about the road; distance; condition
- Main focus of the description or/and hotspot
- GPS coordinates

Summary and highlights.

Detailed description.

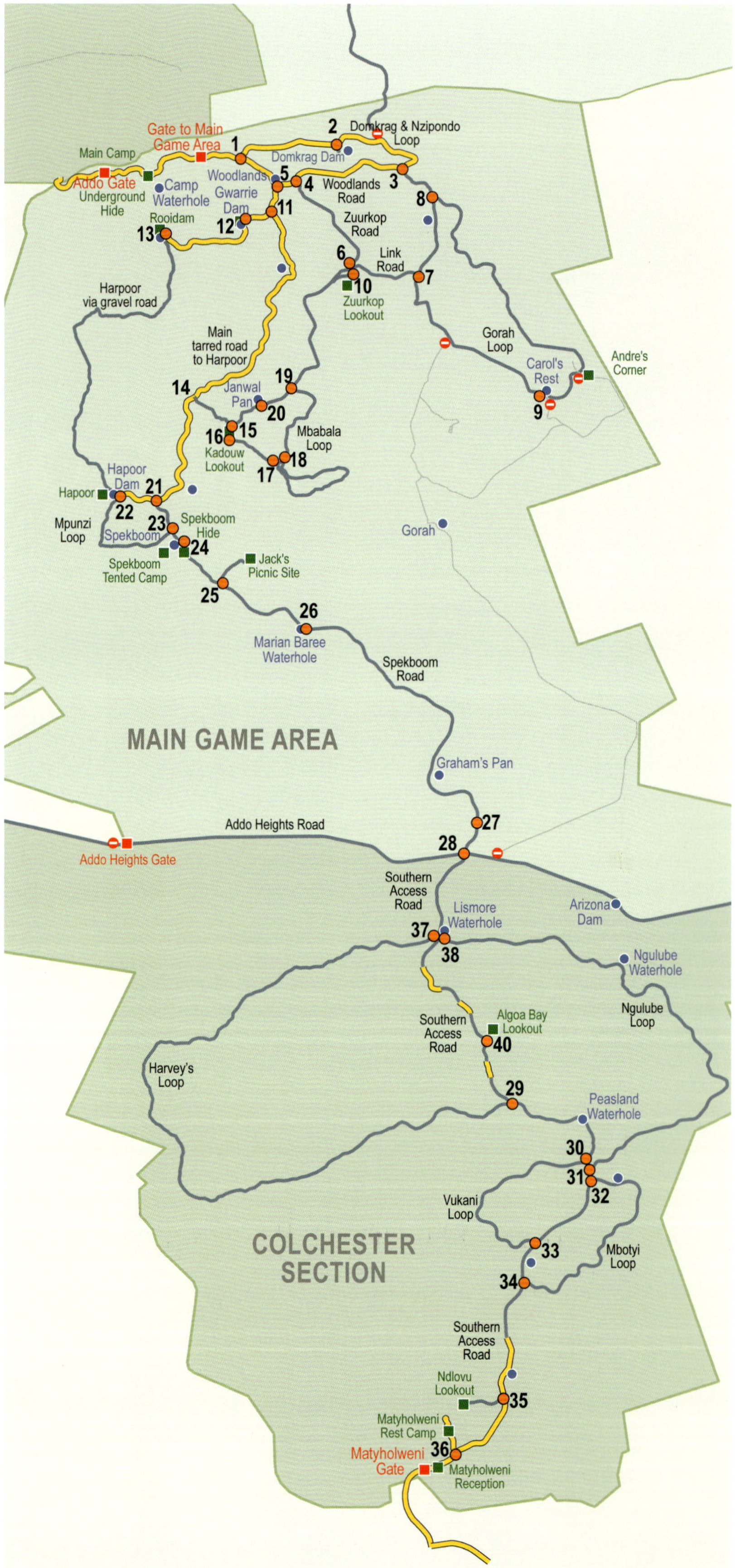

Bokmakierie (*Telophorus zeylonus*)

★★★

Access Road from Main Camp Gate

Main Game Area

From gate to point 1

0.8 km from gate to first intersection, tarred

Main focus: Armstrong fence; Subtropical Thicket

Gate GPS: S33°26'30" E25°45'2"

As you enter the park at the Main Gate, you can look out for game immediately. You often see the largest animal of all munching away on a spekboom or more likely, a smaller one, like a scrub hare, scurrying away into the bush. The access road soon forks out into the Woodlands Road that carries on straight and Domkrag Road to the left. Even though it is a short stretch, it is often surprisingly productive, especially in the late afternoon on your way back when you are rushing to be in time for gate closure. Start off by getting familiar with the Subtropical Thicket vegetation. Plant spotting can be a lot of fun and will enhance your game viewing.

Fencing in the elephants

While waiting for the officer at the gate to check your permit, note the **heavy fence** surrounding the game area. Soon after the park was proclaimed, the authorities were continuously challenged with elephants breaking out and causing havoc in the surrounding farmlands. Some were even killed in collisions with trains. The first park manager, **Stephen Harold Trollope**, had to make a plan. He started supplying the elephants with oranges, hay, pumpkins and pineapples to keep them inside the park. Soon visitors were invited to come and watch the elephants from a platform at feeding time. This practice eventually led to opening the park for self-drive visitors in 1978.

The feeding plan was, however, not sustainable and caused problems. Until recently, visitors were warned not to carry citrus in their vehicles since some of the older elephants still craved the taste and would follow the smell, frightening visitors by following their vehicles. The practice of feeding elephants continued until 1979 before it was completely phased out.

Years later, when **Graham Armstrong** was appointed park manager, he developed an elephant-proof fence that is still in use at some places in the park. The **Armstrong fence**, as it is called, was constructed from tram rails and lift-shaft cables that made it an extremely effective elephant-proof barrier. Marvel at the hard work that must have gone into constructing this ingenious type of fence as you pass through the gate. The modern fences are predator-proof electric fences.

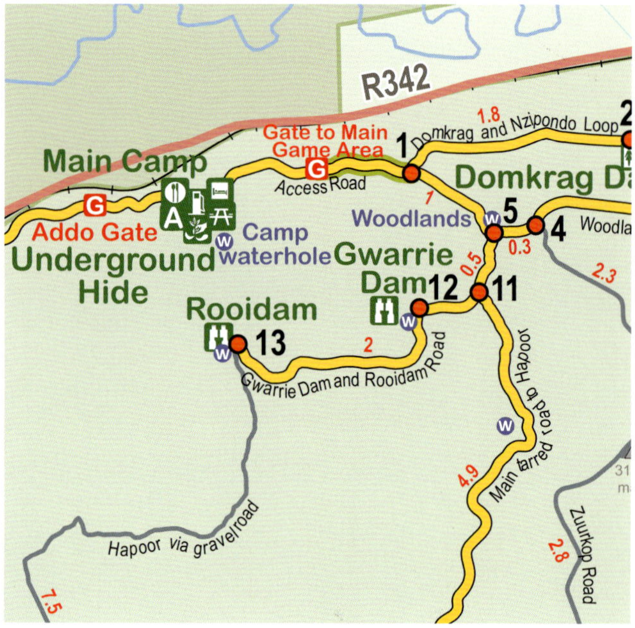

A special kind of vegetation

As you drive into the park, notice the vegetation on either side of the road. This type of vegetation is extraordinary and is called **Subtropical Thicket**. Thicket is not a single plant, but a community of different plants that are intertwined and interdependent. This is the dominant vegetation type within the game-drive area and forms a significant part of the unique **thicket ecological unit**. The particular climate and its animal composition are part of the ecological unit, which is called a **biome**. Addo is the only national park that has been specially proclaimed to conserve the Subtropical Thicket biome.

A wide range of growth forms, including all kinds of succulents, deciduous and semi-deciduous shrubs, dwarf shrubs, geophytes, annuals and grasses contribute to the plant diversity of this vegetation type. Overall it is dense, woody, semi-succulent and thorny, and in its pristine state, it is almost impenetrable.

The **spekboom** is the most extraordinary and conspicuous plant in thicket vegetation. Its name is derived from the Afrikaans word 'spek' which is the word for blubber or bacon and refers to the succulent nature of the leaves. Its scientific name is *Portulacaria afra.* The succulent leaves are bright green, fleshy and egg-shaped. As the tree ages, the colour turns yellowish. In spring after the first rains, it is covered with masses of tiny pink flowers. Spekboom is also called **elephant food** because of the excellent fodder quality for browsers.

According to The Spekboom Foundation, the plant has enormous **carbon-storing** capabilities. Its capacity to offset harmful carbon emissions is compared to that of moist, subtropical forests. This remarkable plant is unique in its capability to store solar energy and perform photosynthesis at night. This makes a spekboom thicket 10 times more effective per hectare at carbon fixing than any tropical rainforest. Each hectare of spekboom could capture 4.2 tonnes of carbon yearly. Yet, outside the park, thicket vegetation is still being cleared to make place for agriculture. It will take up to

150 years or more to regrow to its former climax state.

Scramblers (climbers with thorns) and **shrubs** are common and among these **plumbago** and **Cape honeysuckle** are particularly attractive. Like many thicket plants, they can survive in drought conditions. Sunbirds are attracted to the honeysuckle flowers for their nectar.

Aloes are typical of the Eastern Cape and also occur within thicket. These succulents with ferociously prickly leaves and flowering candelabras rise above the thicket, and their orange to red flowers are visited by insects and birds for the nectar. A beautiful specimen of the **curved aloe** can be seen at the entrance of the dining area at the Main Camp.

Another jewel of the thicket is the **Karoo boer-bean** with its blood-red showy flowers that appear from August to November. The flowers are filled with nectar and attract a wide variety of birds and insects. The small evergreen tree is hardy, much branched and gnarled, with characterful twisted, pale branches and a charcoal-grey trunk. It has fine, feathery but leathery leaves with small leaflets. The seeds are edible and a favourite with many animals, including vervet monkeys.

Marvel at the thicket vegetation as you drive along; **it is extremely special and covers only two percent of South Africa's land surface**. It is also known as Albany Thicket or simply Addo bush with no fewer than 1 558 plant species of which 20% are endemic. For each of these plant species, there are at least 35 organisms that depend on it. The land covered by thicket is semi-arid (rainfall between 200–600 mm) and rain occurs throughout the year as the area falls within the interface of summer and winter rainfall climates.

Brown Hyena (*Parahyaena brunnea*)

★★★★

Access Road from Matyholweni Gate northwards

Colchester Section

Gate, 36, 35, 34, 33, 32, 31, 30, 29, 40, 38, 37 to 28

18.2 km gravel, but tar on gradients

Peaseland and Lismore Waterholes,
Algoa Bay and Ndlovu Lookouts

From south to north or 36 to 28

Gate: S33°40'57" E25°47'44"
Turn-off from N2 to the gate: S33°41'51" E25°48'38"

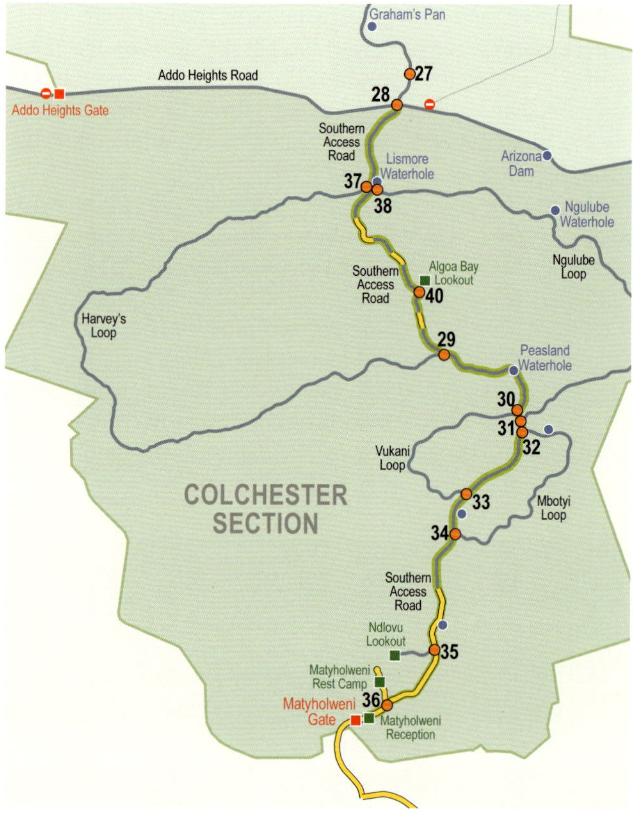

The access road from the south into the park is particularly scenic with beautiful vistas, hills and valleys. The thicket is not over-utilised and in good condition. Plant species diversity is at its prime and game viewing is overall good although game distribution is sparser than in the north of the park. A huge section up to Addo Heights is gravel but the first stretch is tarred and so are sections where there are steep gradients. The gravel sections are rough with occasional potholes, especially in the valleys, making the road unsuitable for buses and towing ordinary caravans. Only 4x4 caravans and trailers are allowed to enter at this gate. When the roads are wet, the gravel patches can become slippery and dangerous. Tarring the entire road is on the cards for the future. Overall it is a beautiful and scenic road, offering good game viewing, plant spotting as well as birding and therefore gets a high rating.

The Colchester Section

The Colchester Section of the park (i.e. the southern section up to Addo Heights) was formed through the purchase of the Olifantsplaat and Vetmaakvlakte Farms. It encompasses 13 000 ha and, together with the Main Camp Area, forms a single **ecological functional unit of 26 000 ha**. The fence separating the two areas was removed in 2006. The area has natural water pans dug out by the elephants over thousands of years. The thicket here is still pristine in places, but remnants of farming activities will remain evident for another century or two. The rehabilitation of purchased land includes removal of old farm fences and structures (e.g. buildings and reservoirs), repair of gravel roads, fencing of the new areas and removal of alien plants.

You may notice bee-boxes erected in strategic places, especially in the valleys where trees tower above the shrub layer. **Elephants love big trees for shade** but also favour them for their leaves and branches for food. To protect special trees from over-utilisation and damage by elephants, bees are used as a deterrent to keep them away. Elephants don't like bees because they usually attack their ears that are particularly sensitive because of the thin layer of skin covering them. They also sting them around the eyes and the tip of the trunk. Elephants bypass these protected trees simply because they avoid places where there are bees.

The bee deterrent works well for elephants, but attracts **badgers**. These skilled and determined animals attack and destroy hives to get to the honey and bee larvae. They are excellent climbers and are tough and fearless. Although they are nocturnal, you may be lucky to encounter one during the day.

Most game animals, including elephants, are dependent on drinking water at regular intervals and tend to congregate in the vicinity of the supply. The result is often over-utilisation and trampling of the vegetation surrounding water sources (take note of this at Hapoor Dam). The Colchester Section forms an integral part of the water gradient for elephants. Its water supply is carefully regulated and has been purposefully kept short to allow a non-homogenous landscape to develop. Under normal conditions it gets drier further away from a river or natural water source and as the supply dries up the animals are forced to move away in search of water. In the case of the Colchester Section, several pans form after rains, thus attracting elephants to disperse over the entire area in search of better feeding areas while remaining close to a water supply. Once the pans dry up, the elephants are forced to move back to places where there is a permanent supply. The result is a healthy seasonal use of the park.

Ndlovu Lookout

Red-throated Wryneck (*Jynx ruficollis*)

Female Bushbuck (*Tragelaphus scriptus*)

Gate, point 36, 35, 3, 33, 32, 31 to 30

The Matyholweni Entrance Gate is a good choice for day visitors but take note that the permit office only opens at 07:00 daily. The turn-off from the N2 is close to the small town of Colchester. While at the gate and on foot, take time to look at the vegetation and identify some of the plants.

The **Matyholweni Camp** is relatively close to the gate and offers accommodation in secluded chalets for two or more people. The setting is perfect, with thicket surrounding each chalet and forming a private wild garden to watch birds, small mammals and reptiles that often visit the camp. A caravan and camping area are also on the cards for the future.

At point 35 turn off to the **Ndlovu Lookout**. 'Ndlovu' is the word for elephant in the Xhosa language. Here you may exit your vehicle at your own risk. This lookout offers magnificent views over the valley below. Note the game path meandering along the bottom of the valley. You may even spot an elephant or other animals using the path far below.

Look out for **bushbuck** on the way out and all along the main road to the north. These are shy antelope that are often missed when driving too fast. Although they are rather large, they keep to dense thicket and seldom move out onto the road. The bushbuck is related to the kudu but the typical spiral horns are much shorter. A bushbuck sighting is special and worth finding. They are browsers but also feed on seeds, fruits, flowers and tender green grass. This solitary antelope is neither territorial, nor does it defend its home range. The stripes and spots on its coat help with camouflage, blending it perfectly with the dappled shade of its surroundings. Cornered bushbuck males can be extremely dangerous, using their sharp horns and hooves to good effect. An enraged bushbuck can even scare off leopard.

Along the main road, especially near waterholes, you are likely to see **vervet monkeys** at play in tree canopies or low hanging branches. These primates live in troops or family groups and feed on plant material such as the flowers and tasteful pods of the **Karoo boer-bean**. They also eat insects, lizards, bird's eggs and nestlings. The youngsters are entertaining when at play, chasing each other up and down trees, hanging from branches and teasing one another. The adults may be involved in grooming, which is a way of cleaning and neatening the fur, getting rid of ticks, scabs, flakes of skin and salty deposits caused by perspiration. These activities all serve to secure bonds and form alliances between individuals, and reinforce hierarchies.

From point 30, 29, 40 to 38

The stretch of road between point 30 and 29 is usually quite productive. It passes the **Peaseland Waterhole** where you are sure to find some or other kind of game.

A time ago, one of the park giants died close to the solar pump on the far eastern side of the road. The result was a big congregation of all the hyena in the park. During the day **spotted hyena** and the ever-present **black-backed jackal** took centre stage, while **brown hyena** probably skulked around for their part of the bounty under the cover of darkness. Imagine how these scavengers must have competed to rip the huge elephant apart.

Hyena were frantic to sever the bones off the limbs to carry away. Soon all that was left were parts of the huge skull and other big bones, too large to lug out of reach of others. Jackal were the biggest irritation, trying to steal away the bounty from the successful hyena. Gorging themselves and competing with others mean hard work. Every now and again a hyena would trot away towards the waterhole on the other side of the road for a cooling splash and some water. It only took the hyena and jackal a few days to finish the carcass and soon they took off, each in another direction, either back to their dens to regurgitate the food for the young or to rest before going on their next outing to find more food.

The presence of so many carnivores scared away other game. A herd of **red hartebeest** remained at a distance and watched as a **spotted hyena** circled them from behind. Somehow, they sensed that he was not on the hunt, otherwise they would have run. In Addo the spotted hyena are not only scavengers, but also hunt themselves. Red hartebeest are extremely swift runners that can reach speeds of up to 65 km/h and have the stamina to maintain this over a long distance. Red hartebeest are pure grazers and can live without water, as long as they can derive their required moisture from their food.

It is interesting to note that in a place like Addo Elephant National Park, vultures are absent. Vultures play such an important role in cleaning up the environment, but sadly these essential birds are on the brink of extinction and in Addo, they aren't seen anymore. Their role is now taken over by crows. You will notice the presence of three different species; the **pied crow**, the **Cape crow** and the **white-necked raven**.

Point 38 to 28

Once you pass point 29, the road ascends one of the many hills. The **views** along here are magnificent. Note the difference in vegetation as you reach the top of the hill. Be on the lookout for the sign indicating the **Algoa Lookout** point on your right. The signpost is not very clear and is easily overlooked. From this lookout you can see far into the distance, marvel at the rolling hills of the Addo environment, find the coastline and dunefields beyond Colchester and look deep into the Algoa marine area, which is also part of the Addo Elephant National Park.

You are allowed to leave your vehicle at your own risk, but observe carefully before you get out and stay in the demarcated area. Note the vegetation. After rains the spring flowers here are a delight.

The intersection between points 37 and 38 is a hotspot for game. Look out for **buffalo** that are often encountered here. They need to drink water daily and bachelor bulls or mixed herds often use **Lismore** or **Marion Baree** Waterhole

Blackbacked Jackal (*Canis mesomelas*) and Spotted Hyena (*Crocuta crocuta*)

for drinking. Mud-wallowing is an important way for them to regulate their body temperature since their huge bodies can weigh up to 800 kg and generate a lot of heat. During the heat of the day they need shade and rest while they ruminate. Buffalo are **bulk grazers** but during dry periods when grass is scarce, they will also browse. These bovines are attracted to fresh young grass but will also readily feed on old grass. By doing so, the old grass is trampled, which in turn stimulates new growth.

When the park was established in 1931, there were only 30 buffalo left from the great herds that originally roamed the area. The **buffalo** in Addo are the biggest, **truly disease-free** herds in South Africa and therefore highly valued. The revenue generated by the sales of some of these disease-free buffalo has enabled the park to purchase vast tracts of land for conservation. In the early days buffalo were seldom seen during the day. This was probably a behavioural mechanism adopted in the hunting days when hunters were active during the day. The over 400 buffalo are now spotted during the day more often, owing to the influence of lion reintroduction.

Elephants are also regularly seen around here, but the **Lismore Waterhole** is among the waterholes fenced off to deter elephants from drinking there. This is part of an experiment to maintain a water gradient in the park. Elephants are dominant at waterholes. When they are present, other species are prevented from drinking. The aim of the experiment is to **encourage the elephants to find water at other waterholes,** e.g. at Marion Baree.

The experiment involves a rectangular-shaped structure with four corner posts, placed around the waterhole. An electrical strand is suspended at two metres in height and has vertical, one-metre strands hanging from the main strand at 800mm intervals. These strands are electrified to stop elephants from walking in. They soon realise the water point is not available to them and proceed to other water sources that are able to sustain them. Other game is then able to use these water points without having to compete with elephants.

From Lismore the road ascends to Addo Heights. Notice the tall-growing euphorbias on the rocky hillside to the left of the road as you approach Addo Heights. A few specimens also grow on the righthand side of the road. This **tree euphorbia** is endemic to southern Africa. Similar to all euphorbias, it has a white milky latex that is highly toxic, as well as being a severe irritant.

Lion (*Panthera leo*)

Tree Euphorbia (*Euphorbia triangularis*)

★★★★★

Domkrag Road including Nzipondo Loop

Main Game Area

- 1, 2 to 3
- 1.9 km, tarred
- Domkrag Dam
- Domkrag: S33°26'25" E33°46'57'

There is always something interesting to see along this loop and it is a big favourite for an afternoon drive from Main Camp. Initially the thicket is dense but then it becomes more open as you approach the low hilltop towards the dam. Shrubs become smaller and sparser while grass and low karroid vegetation greatly improves visibility into the distance. Birding and game viewing are excellent because of good visibility. The road is tarred and easy to navigate and the Domkrag Dam is a 'must' for any visit. This road has the highest rating.

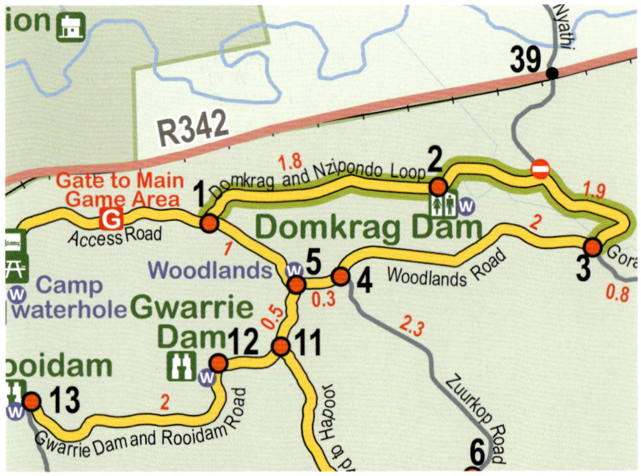

Leopard Tortoise (*Psammobates oculifer*)

South African Shelduck (*Tadorna cana*)

Point 1 to 2

Domkrag Waterhole was named after a spirited **leopard tortoise** that roamed the Addo Park for 45 years. This huge tortoise had the extraordinary habit of getting under parked cars and lifting them up like a car jack (*domkrag* means car jack in Afrikaans). At an estimated mass of 45–60 kg, he was a giant that did not tolerate any rivals. Perhaps he regarded cars as rivals? Adult males of this species usually attain a mass of about 20 kg and when Domkrag came across one of his rivals, he simply wedged his shell underneath the shell of the rival in tortoise style and flipped him onto his back. This would leave the rival helpless with legs waving in the air. He died in 1976 after he fell into an aardvark hole and could not get out. Perhaps he landed on his back? You can still see his shell in the **Interpretive Centre** at Main Camp.

Look out for tortoises as you drive along. These reptiles are often called '**padlopers**' – literally translated as 'road walkers' from Afrikaans – because of their habit of strolling along roads. Tortoises are numerous in the park and regularly drink at waterholes, but in winter they hibernate and are therefore mostly seen in the warmer seasons.

It is interesting to note that palaeontologists think tortoises evolved in the Karoo Basin and that the first tortoise 'ancestor' was a **bradysaurus** that lived 265 million years ago. It was as large as the Volkswagen Beetle. South Africa has five species of tortoises and several subspecies; a higher diversity than anywhere else in the world.

Scrub Hare (*Lepus saxatilis*)

Warthogs love the type of vegetation found along this loop. The open patches with low vegetation suit their lifestyle well. Watch a warthog family and observe how they graze. Observe how they use their tough, muscular snouts to dig up roots and rhizomes and how they go down onto their knees to get better leverage. Males have two prominent sets of warts on the face while females have only two distinctive upper warts. These warts are not unusual growths like true warts, but are deliberate growths on the skin used to protect the eyes and face from injury by the inward growing tusks when fighting. The canine 'tusks' in warthogs should be called **tushes** because they differ from those of elephants. Both are ivory but in the case of elephants, the tusks are incisors and not canines. In warthogs the upper canine tushes curve out and upward and can become very large in males. Take photographs of males you encounter and compare them later to see which had the most impressive tushes. Notice the ridge of white hairs along the sides of the lower jaw that is more pronounced in females.

You can alight (at your own risk) at the platform overlooking the **Domkrag Dam**. Use your binoculars to find the numerous **waterbirds** in and around the dam. The Main Game Area of Addo has no perennial river and dams and pans (after rain) are the only other water bodies. In spite of this, Addo

Pale Chanting Goshawk (*Melierax metabates*)

has a surprising variety of aquatic birds. The **duck family** is represented by geese, teals, ducks and shellducks. Birds **foraging in shallow water** include the herons, hamerkop, sacred ibis, moorhen, plovers and lapwings. Grebes, coots and cormorants are **swimmers** and **divers**. See how many of these waterbirds you can find.

In the bushes around the platform you can expect to see black-capped bulbuls, southern masked weavers, Cape weavers and the beautiful two songsters, the boubou shrike and bokmakierie.

Point 2 to 3

This stretch of road is called the **Nzipondo Loop**, probably because **springhare** are often seen here during night drives. 'Nzipondo' is the Xhosa name for springhare. This kangaroo-like animal is neither a hare nor a rabbit, but a rodent. The veld around Nzipondo Loop is ideal habitat for them since they prefer open places with sandy soil to dig their burrows and hide during the day. 'Spring' means jump in Afrikaans and that is exactly what they do best.

The **pale chanting goshawk** is often spotted in this vicinity. This is a widespread raptor, mostly confined to dry areas. It feeds on a variety of prey items but in Addo its main prey seems to be the abundant **Karoo rat**. The raptor derives its name from its characteristic chanting call. Look out for the diurnal Karoo rat often seen dashing from one shrub to the next.

This is a good area to find the **biggest bird in the world**. The vegetation type is suitable for the flightless **ostrich** which feeds on forbes, grasses, seeds and other vegetable matter.

The **largest land mammal** in the world, the **elephant**, is regularly seen around here when making its way to the Domkrag Dam or nibbling here and there on juicy vegetation after drinking.

Because of the large open spaces, visibility is excellent and the **biggest predator** in Africa (the lion) and other predators can easily be spotted when in the vicinity. The shy **eland** is the **biggest antelope** in the world and is also seen along this route from time to time. Perhaps we should call this the route of giants?

Malachite Sunbird (*Nectarinia famosa*)

★★★★

Gorah Loop

Main Game Area

- 3, 8, 9, 7, 8 to 3
- 12.7 km, gravel
- Carol's Rest Waterhole, Gorah Kloof/Gorge, grassland areas
- Carol's Rest: S 33°29'3" E25°49'16"

Gorah Loop is one of the favourite drives because of the good visibility in places, the species diversity, the intermittent grasslands (remnants of farming activities), the waterhole and the scenic aspects. These secure a good rating; however, the road is slippery when wet and therefore often closed after rains. Red or greenish-grey mudstone underlies most of the Addo Park. The mudstone weathers to form clay and this type of soil is also easily eroded.

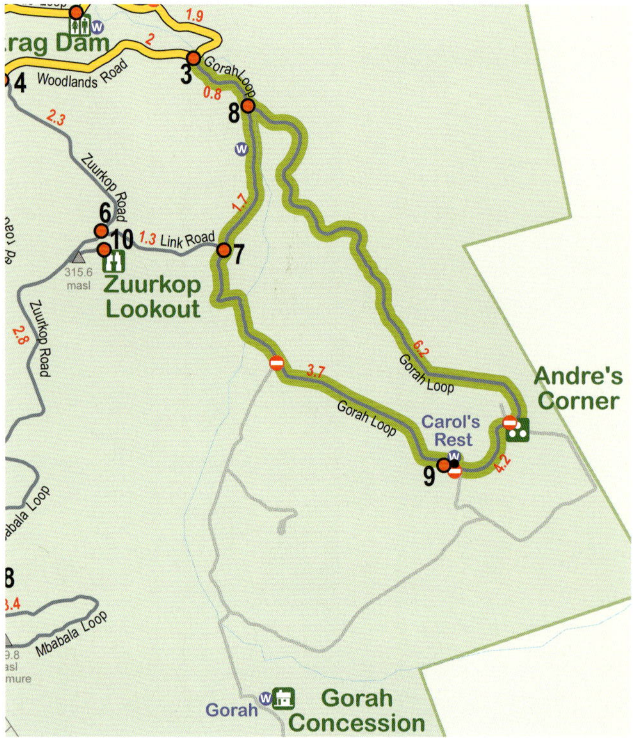

Black Rhinoceros (*Diceros bicornis bicornis*)

From point 3 to 8

The Gorah Road has an interesting history. In **1992** the **farm, Gorah**, was included into the park. The elephants made a rush for the area because of the abundance of prickly pears (an invasive plant), which happens to be a favourite of theirs. Tourists complained they were not seeing any elephants. Consequently park officials had to construct a new tourist road into the Gorah area as the elephants were no longer visible to the tourists in the original section.

A huge part of the original Gorah Farm, with the rebuilt **Gorah mansion**, **is now a smart lodge** within a concession area south of Carol's Rest Waterhole.

The elephants were a big help in eradicating the **invasive prickly pears** in that area. For a long time after all the prickly pears were cleared, the elephants still preferred to browse the thicket along Gorah Loop and drink at Carol's Rest Waterhole. As you drive along this loop you will notice that the thicket, which was once dense and lush, is now **over-utilised and looks tired**.

Officials had to make a plan to get the elephants to explore other parts of the park. This is when they started experimenting with ways of getting elephants to move to other feeding grounds – not an easy task because elephants are intelligent, have an excellent memory and don't allow themselves to be herded. In addition, about a third of their diet consists of the couch grass that is available in abundance on the intermittent grasslands along this loop.

From point 8 to 9

A short distance from the entrance to the Gorah loop, the road splits. **Take the left turn** and drive through a slight depression. This depression is the end of the **Gorah Gorge**, a kloof that starts further south near the original Gorah homestead where there is a **natural spring**. Even the earliest inhabitants of this area relied on this spring known as 'de Gorahs'; the place of natural water. In the 1800s the area was colonised by a farming family but at the turn of the century, during the Anglo Boer War, the homestead and farm were left in ruins. After the war the farmhouse was rebuilt and later, in 1992, incorporated into the Addo Elephant National Park. A number of sites featuring stone implements are located throughout the park. The **Gorah farmhouse is now a national monument** and has been restored to its former glory. Evidence of a **Khoisan settlement** can still be found in the area behind the Gorah house.

After the first incline, the bushes open up onto a **big grassy plain** where the bush was cleared for agricultural purposes long before this area was incorporated into the park. This plain is now slowly reverting back to the original vegetation, but it will take a century or two (if ever) before it is completely restored to its original state.

In the meantime, the grasslands **provide grazing** for the grazers, but also for the elephants. Here you may encounter warthog, red hartebeest, eland and zebra. Kudu are not natural grazers but are often seen crossing these grasslands and nibbling on herbs and forbes along the way.

When the park was proclaimed, not only elephants were almost eradicated, but other game too. Game translocations only became possible much later, towards the end of the 1900s. From written records by early pioneers it could be established which species occurred naturally in the Addo region before commercial hunting started. This enabled SANParks to restock the park with game that geographically belong here.

In 2003, **six Kalahari lions**, as well as the first **spotted hyena**, were introduced to the park. Some of the present lion pride are often encountered on the Gorah Loop.

Warthogs are abundant today, but **none were left before 1996** when a viable number were translocated from other parks. These hogs play an important role in the ecology. It is the only pig species that has adapted to grazing. During the wet seasons, warthogs graze on short perennial grasses but during the dry season they subsist on bulbs, rhizomes and nutritious roots. Although they can flee by means of fast sprinting, they are suitable prey for lion, spotted hyena and leopard, among others.

Gorah Loop is not only a hotspot for grazers, but **grassland and terrestrial birds** are seen here in abundance. The open plains make birding easy, but grassland and terrestrial birds do not like a close approach. The best way to observe them is to patiently wait for foraging birds to move closer.

Denham's Bustard (*Neotis denhami*)

The larger terrestrial birds such as **secretarybirds**, **cranes** and **bustards** are nomadic to some extent and are often totally absent. The biggest of them all, **ostriches**, are present most of the time and can be observed parading the plains to feed on greenery, rhizomes and grass, including grass seeds that are deftly stripped off the stem.

There are not many places where **Denham's** and **Ludwig's bustards** occur in the same area. In Addo, however, they sometimes even forage together. These huge birds are mostly very skittish and must be approached slowly. The males of both species perform a spectacular display during the breeding period.

The presence and numbers of **blue cranes** fluctuate depending on rainfall patterns. These graceful birds also perform a striking dancing display during which one of the birds will present something like dung or plant material to its partner.

Smaller grassland birds are regularly seen and these include crowned lapwings, ant-eating chats, African pipit and capped wheatears, spotted thick-knees and rufous-naped larks.

There are no vultures in the Addo region to help clean up carcasses and **crows and ravens** have taken over this role to an extent. It is therefore not surprising that all three members of this family occur here. They include the white-necked raven with its huge white-tipped bill, the overall glossy black Cape crow and the pied crow with its white upper back and upper belly.

Carol's Rest Waterhole

The waterhole at the far end of Gorah Loop is among those **closed to elephants**. This means that all kinds of smaller game are regularly seen drinking here. The waterhole was named after the wife of an earlier game ranger who had a special attraction to this waterhole. She was killed in an accident and the place was named in her honour.

Vervet monkeys are mostly seen close to water. Their antics can be very entertaining and watching them at the waterhole from your vehicle can be a rewarding experience. You may even be lucky enough to see some of the shy inhabitants of the park around here. **Kudu** are everywhere, but its close relative, the **eland**, may also be around the area. Another shy inhabitant that may surprise you anywhere in the park, is the **black rhino**. Scan the edges of the thicket to increase your chances.

Blue Crane (*Grus paradisea*)

Towards Gorah Gorge and beyond

From Carol's Rest the road passes through more heavily grazed thicket before it descends into the shallow gorge. Look out for **guinea fowl** usually seen around here. Notice the difference in appearance of the vegetation down at the bottom of the gorge. This is usually referred to as Valley Thicket and is dominated by **sweet-thorn acacia**.

Look out for **scrub hare** around here and as the road ascends to the open plains before the Link Road, you are likely to see **yellow mongoose** and even **suricate** foraging in the veld.

Small herds of **red hartebeest** are a common sight on these open plains of Gorah Loop. The red hartebeest is an ungainly-looking antelope, with a long narrow face, staring goat-like eyes with a tuft of hair under each one and crooked horns. Both sexes look alike, with the males being sturdier built and sporting larger horns. If your visit coincides with the calving months of October to December, you may initially see no calves around since the mother hides them for a while before she allows the youngsters to join her while grazing. The small calves that you may see suckling will already be a few weeks old. It takes eight months until they are weaned, and then they form groups with others of their age in the herd.

Plains zebra were reintroduced back into the park in 1996. Zebra play a critical role in the overall ecology because they remove old grass growth and consume lower quality plant matter. This enables new shoots to grow and increases the overall quality of vegetation for other herbivores in areas where they graze. You will surely encounter a herd or two of

Plains Zebra (*Equus quagga burchellii*)

zebra on the Gorah plains. It is easy to spot them for they usually graze in the open. Their main predators are lion and spotted hyena.

Zebras are highly social and usually form small family groups called **harems**, which consist of a single stallion, several mares and their recent offspring. Bachelor groups that are all-male also occur and have an age-based hierarchy led by a young male. The males stay in their groups until they are ready to start a harem. The bachelors prepare for their adult roles with **play fights** and **greeting** or **challenge rituals**. Stallions form and expand their harems by abducting young mares from their natal harems.

When a mare reaches sexual maturity and has her first oestrus cycle, she attracts the attention of nearby stallions, both bachelors and harem leaders. Her family stallion (likely her father) chases off or fights stallions attempting to abduct her. Even after a young mare is isolated from her natal harem, the fight over her continues until her oestrus cycle is over and it starts again with the next oestrus cycle. The male that impregnates her keeps her for good and she becomes part of his harem.

Mares also exist in a hierarchy, with the alpha female being the first to mate with the stallion as well as the one who leads the group. When new mares are added to the group, they are met with hostility by the other mares. Thus, the stallion must shield the new mares until the aggression subsides. The most recently added females rank lowest or females that become unfit or weak may drop in their rank. The female membership of a harem stays intact even if a new stallion takes over. A stallion defends his group from other males.

Warthog (Phacochoerus africanus)

Orange-throated Longclaw (*Macronyx capensis*)

Helmeted Guineafowl (*Numida meleagris*)

★★★★★

Gwarrie Waterhole and Rooidam

Main Game Area

11, 12, 13

2 km, tarred

Gwarrie Dam, Rooidam, elephants, hyena

Gwarrie Dam: S33°27'10" E25°45'51"
Rooidam: S33°27'16" E25°45'2"

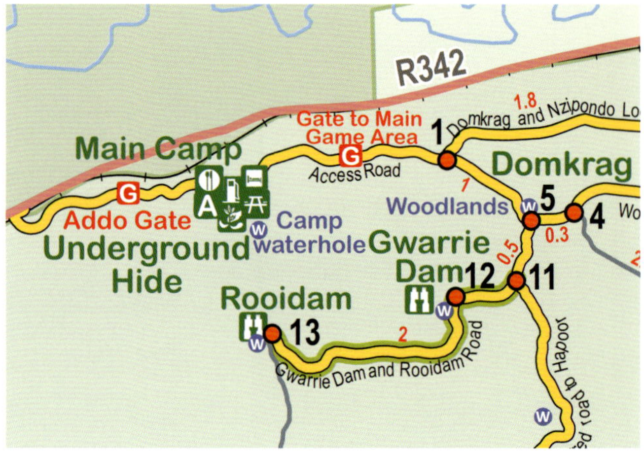

This is a popular destination for a midmorning or afternoon drive. A tarred road leads to two popular, permanent dams on the western side of the main road. The Gwarrie Dam used to be a gravel pit and Rooidam has a proper dam wall on the far side. Both dams are surrounded by typical thicket, but the surrounding area at Rooidam already shows signs of erosion. Plenty of parking allows visitors good views over the water. When elephants come drinking they dominate other species, but at Rooidam hyena take no notice of their superiority. Birding is particularly good along the first section of the road. Bokmakieries are as iconic to Addo as the elephants. Look out for these striking beauties with their melodious duet.

Gwarrie Dam

Some years ago, the media carried reports about an **extraordinary drama witnessed at Gwarrie Dam**.

A breeding herd of elephants with yearlings and subadults approached the dam to drink. One of the cows was protecting a small baby, keeping it close and touching it from time to time with her trunk. It must have been only a few months old because it was still a bit wobbly on its feet and fitted comfortably under the mother's tummy. While the mother and the rest of the herd where quenching their thirst, baby was exercising its little trunk and mimicking the elders. The next moment it lost its balance and slipped into the water. The mother and the rest of the herd were immediately attentive and tried to help it to get out, but it kept on slipping back. The herd tried to wet and crumble the embankment, but to no avail. The little one was getting anxious and the elders stressed out.

Park rangers were called for help. They tried their best to break part of the embankment for the baby to get out, but this agitated the herd even more, making them aggressive towards the rangers. The only plan the rangers could think of was to get the herd out of the way to allow them to rescue the calf. Thankfully, the plan worked. They used the vehicle to herd the concerned adults out of sight while the rescue team jumped into the water to get the baby out. This was not an easy task since the baby resisted and must have weighed more than its birthweight of 120 kg. It kept slipping back and to heave it out against the slippery edge was hard work.

Once the baby was out, they had to get it back to its mother, which was now nowhere to be seen. The only thing the rangers could do was to load the baby on the bakkie and find the stressed natal herd who by this time had moved deeper into the bush. After a three-hour ordeal, the family had been safely reunited, thanks to the bravery and dedication of SANParks rangers in Addo.

Young Elephant bull bathing

Guides in Addo don't carry firearms because the Addo elephants are some of the gentlest in the country. They don't feel intimidated around vehicles and even bulls in musth pose little threat. That said, tourists should never get too close to any of the elephants as it makes them feel uncomfortable; rather give them space and park the car a little further away. More often than not, the elephants will come to you.

The Gwarrie Dam is not a particular favourite with elephants. It may be that it is too deep and the embankment too hard. A white calcrete deposit on the banks gives an indication of the high concentration of calcium carbonate in the soil. This may also affect the water quality.

Be aware that there are signs of excitement to look out for when close to elephants. The temporal gland is situated under the skin between the eye and the ear and has a tiny opening on the temple. When an elephant gets excited the gland secretes an oily substance, which may leak from the opening and stain the skin. Excitement can be caused by various factors such as irritation, anger, group interaction and pleasure. When bulls are on heat, it is referred to as being in musth. Bulls secreting copiously from the temporal glands together with constant urine dribbling, are probably in musth. Be careful if you see any of these signs when you are close to elephants.

Baby Elephants at play

Elephants in greeting ritual

Rooidam

The soil around this dam is a reddish brick colour and *rooi* is the word for red in Afrikaans. This dam is a big favourite on hot summer days when the heat drives elephants to the water to cool off.

Large animals such as elephant, rhino and buffalo lose heat relatively slower than small animals because their surface area is smaller in proportion to their body mass. This is why they tend to find ways of lowering their body temperature, by either seeking shade or visiting the dam to have a good wallow or swim.

Elephants are at home in water and can swim extremely well. They enjoy playing and frolicking, sometimes completely submerging with only the tip of the trunk out of the water. Wallowing in muddy water is the equivalent of a spa treatment for them. They froth and kick up muddy water, churning it into a thick consistency, and then slap it onto their skin to act as a protective mud pack against the hot sun and parasites.

After bathing or wallowing, elephants often cover themselves with dust when they are resting. If no loose dust is in sight, they will soften the soil with their front feet. Once enough soil can be scooped up with their trunks, they use it to dust both over and under their bodies.

One of the spotted hyena clans of the park den in the vicinity of this dam. They keep out of elephants' way and are usually seen early in the morning. Once, a member of the clan hid a severed bone under the embankment on the side of the parking lot. It must have been really smelly by the time he collected his quarry, but to him it was like pudding. He pulled it out of the water and deftly ran up the hill in the opposite direction of the den; probably to keep it to himself. A younger member of the clan tried to follow, but quickly gave up when he sensed that he wasn't welcome.

Spotted Hyena (*Crocuta crocuta*) at Rooidam

Bathing Elephants completely submerged but for their trunk tips

Young Elephants playing in the water

Having great fun wrestling

★★★★

Hapoor Dam via gravel road

Main Game Area

- 13 to 22 (includes Hapoor)
- 7.5 km, gravel
- Hapoor Dam, elephants
- S33°30'10" E25°44'24"

A gravel road continues from Rooidam to Hapoor Dam. For most of the way visibility is restricted to a few metres into the thicket, but occasionally it is possible to see further afield. The species diversity is not great but expect to see browsers such as kudu and lots of elephants. Occasionally mongoose, zebra, warthog, buffalo and hyena may be seen. Birding is slow. The road used to be a narrow one-way from Hapoor to Rooidam, but it has now been upgraded and widened for two-way traffic. Overall it is a popular road and gets a high rating.

A breeding herd rushing to Hapoor Dam

From point 13 to the Hapoor Dam

Hapoor Dam is the sparkling jewel of elephant sightings in the park. The dam is man-made and its water is supplied by Caesar's Dam situated just outside the southwestern border of the park. Caesar's Dam was built by the Department of Irrigation during the 1920s. This was the time of the great elephant massacre and the area around the dam was dominated by a fierce elephant bull. He was greatly feared and became a legendary figure in the area. As ruler of the region he became known as Caesar and his name was given to the dam. But this was long before the proclamation of the park.

All the elephants in Addo know the **Hapoor Dam** and visit it from time to time. Elephants are intelligent creatures. They know and remember where they can get the best water, where the best feeding grounds are and they pass all this on to their progeny. They also know where the inflow of fresh water to each dam is located and that is where they all want to go. Unfortunately for the lower ranking ones, they have to wait their turn as high-ranking individuals are the first to drink and lower-ranking ones have to yield when they approach. They prefer to drink clean water and will patiently wait their turn at the inflow to a dam rather than drink muddy water.

Visiting a waterhole is the highlight of an elephant's day. Water is drawn through the trunk and transferred to the mouth. The trunk can hold up to 10 litres at a time and they drink between 150 and 200 litres of water daily.

Speckled Mousebird (Colius striatus)

The trunk is the most unique feature of the elephant body and it is used as a dexterous tool. Watch how cleverly they use it. It is an extended union of the upper lip and nose, with the nostrils continuing from the base of the nose all the way to the tip of the trunk as two separate openings which end in two highly sensitive and mobile finger-like extensions. It is composed of 60 000 different muscles, blood vessels and nerves, yet contains no bone or cartilage. The thick skin is sparsely covered in hair and a few bristles and gathers up in wrinkles as the trunk is flexed.

Elephants use the trunk for feeding, watering, dusting, smelling, touching and lifting. When they approach water and are unsure about their safety, they are often seen sniffing the air with their trunks, thus gathering information about their surroundings.

Babies having fun at play

★★★

Hapoor Dam via main tarred road

Main Game Area

- 5, 11, 14, 21, 22
- 11.5 km, tarred
- Subtropical thicket species, kudu
- Point 5 to Hapoor Dam
- Hapoor Dam: S33°30'10" E25°44'24"

Cruising along the shortest route towards Hapoor, it is easy to forget about the speed limit. It is not a particularly scenic road and the thicket can become a monotonous alley when deep in thought, but beware, the gently winding road can hide elephants from sight and wake you up in a jiffy. On the other hand, you may spot a group of elephants in the distance and when you get there seconds later they are gone, swallowed by the thicket vegetation. Although the road is beautifully tarred and lined with dense thicket, game viewing is average because visibility is low and so is the scenic appeal. The road gets an average rating although it leads to the most exciting, five-star waterhole in the park.

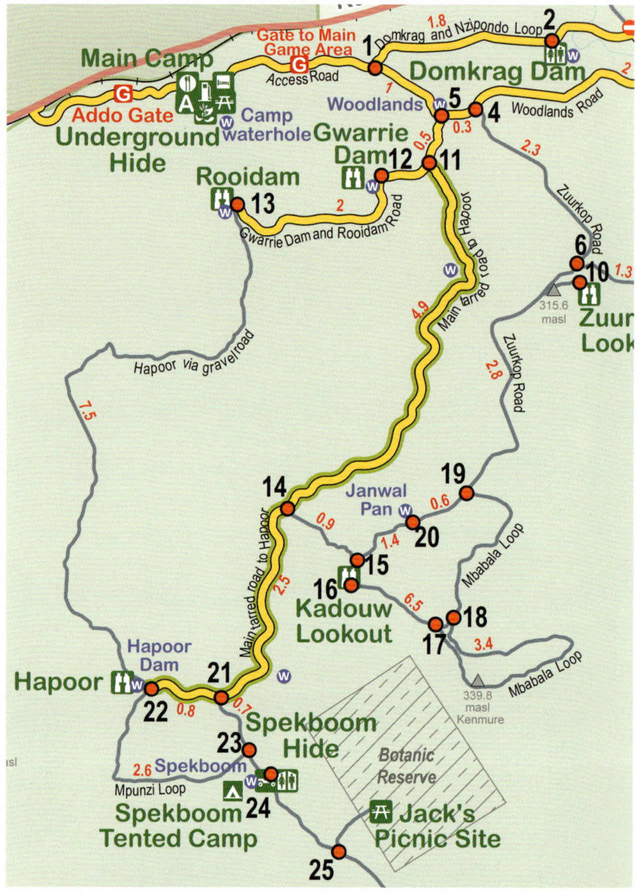

The tarred main road

Along this main tarred road, the Subtropical Thicket averages two to three metres in height and looks impenetrable. In spring, after rains, the **spekboom** bushes burst into flower and the soft pinks and mauves are a delight to the eye.

Antelope like **kudu** thrive in this environment. They have the cover of the thicket, are not affected by the thorns or the dense undergrowth and derive enough moisture from the plants they eat, which means they are not dependent on drinking water during times of drought. We have to wonder how they manage with their massive horns, but adult males simply lift their heads when running through the dense thicket, thereby laying their long spiral horns flat on their shoulders to avoid them getting stuck in branches. The horns of a mature bull kudu have two-and-a-half twists, and if straightened, would reach an average length of 120 cm. However, they may occasionally have three full twists. The horns do not begin to grow until the bull reaches 6–12 months, twisting once at around two years and not reaching the full two-and-a-half twists until the age of six.

Although Addo Elephant National Park does not have a wealth of game birds, it is the only national park in South Africa where **red-necked spurfowl** can regularly be seen. Elsewhere in Africa, red-necked spurfowl have a wide distribution. They prefer the edges of thickets where they forage in pairs for tubers, bulbs, seeds, fruit and invertebrates.

Red-necked Spurfowl (*Pternistis afer*)

Hapoor Dam with breeding herd waiting among the vehicles

Hapoor Dam

The **Hapoor Dam** is a favourite destination with visitors. The wide-open space around the dam is excellent for viewing. There is ample parking close to the water and elephants don't seem to mind the presence of vehicles and often pass right next to them. Despite the relaxed nature of most of the elephants, remember that bulls in musth can turn aggressive and cause an incident. It is better to keep a distance. When elephants approach your parked vehicle, don't panic and don't keep your engine running.

The dam was called after a notorious bull that ruled the park for 24 years. He was called Hapoor and everybody kept out of his way. Hapoor had a strong personality but also an extremely aggressive nature. He was huge and hated humans. This was probably due to a bad experience with a hunter who missed his aim and shot Hapoor through the ear. For the rest of his life he had this nicked ear (In Afrikaans: *hap oor*) that reminded him of his arch enemy.

In his lifetime Hapoor sired most of the calves born between 1944 and 1968 and the Addo elephant herd increased from 20 to more than 50 during his tenure. This, of course, meant the gene pool of this isolated herd was rapidly shrinking. In the meantime a bull called Bellvue was growing into a massive animal. In 1959, he challenged Hapoor and tried to depose him, but Hapoor was too strong for him. They fought a battle to the death, from which Hapoor emerged as the unscathed victor.

In another incident in 1966, an older matriarch called Ouma developed a large abscess and the men from Nature Conservation decided to help her by removing the abscess under sedation. While applying and waiting for the sedative to take effect, Hapoor and a cow called Eentand (translated from Afrikaans to 'one-tooth') suddenly arrived, noticed the unsteady gait of the cow and immediately came to her rescue. Hapoor was visibly concerned and the two elephants tried their best to help Ouma stand up. When she finally collapsed and appeared to be dying, he stabbed her four times with his tusk straight into the brain, probably to relieve her from her suffering.

Two years later, one of Ouma's sons called Lanky, challenged Hapoor for dominancy. Hapoor succumbed to the younger bull and was driven out of the herd to become a loner. Hapoor never conformed to any restrictions and being ousted did not suit him. He consequently broke through the

African Buffalo (*Syncerus caffer*)

Elephant mothers are protective of their young

Mud rolling at the dam is a favourite with young and old

'elephant-proof' Armstrong fence and was the only elephant ever to achieve this.

He was soon found in the region of the Coerney Station. Knowing that no fence would be able to stop him, the warden had no choice but to end his life. In those years it was not yet possible to dart and translocate elephants and there was no way of herding him back because of his aggressive nature.

Longevity in elephants is about 55 to 60 years. Hapoor was a mere 44 years old. Elephant cows are usually longer-lived than bulls, which often become involved in fighting for dominancy.

The Hapoor Dam is named after this famous bull elephant and a cast of his head was made that can be seen in the interpretive centre at Main Camp.

Elephant society

Elephants are **socially complex**. You may have noticed male and female elephants live entirely different and separate lives. **Males are usually alone** or accompanied by one or two other males, mingling with female herds only for the purpose of mating. Males will frequently fight with other bulls to determine dominancy as **only the strongest are able to mate** with receptive females. In this way a strong, healthy progeny is ensured.

The **females and their young** are guided by the oldest and largest female of the herd, the **matriarch**, which presides over an average of **five to 15 adult cows**. The herd is made up of mothers, daughters, sisters and aunts as well as immature males and females. The matriarchal herds maintain close bonds among themselves but also interact with other herds, families and clans. They never forget their family roots and keep track of each other by vocal and non-vocal communication. There are seven family groups in the park. Subgroups may break away from the rest for some time and then rejoin their herd again later.

From the initial herd of 11 elephants that survived the massacre in 1919/1920, the population in Addo has now **grown to more than 600**. Thanks to the advances made in mammal translocation methods in the last decades of the 1900s, it has become possible to translocate mature bull elephants from the Kruger National Park to Addo. Injecting new genes into the Addo gene pool became necessary to ensure the well-being of future generations. Of the eight Kruger bulls, only four remain. Initially they did not breed until they eventually ousted some of the Addo bulls. They have now settled in well and are mating with the Addo cows.

A large majority of the elephant cows in Addo have no tusks. The introduction of bull elephants from Kruger has now widened the genetic pool and therefore tusks are becoming more common.

The Addo Elephant National Park is busy extending the elephant home range. Elephants have already been translocated to the Nyathi, Darlington and Kuzuko areas.

As soon as the elephant-proof fence around 29 000 hectares of land in the Kabouga section is completed, it will allow for the translocation of another large number of elephants across to this area. The fence around the 50 000-hectare Darlington section has been completed and many elephants have already been relocated there.

Mud bathing is important for getting rid of parasites and to cool off

★★★★

Harvey's Loop

Colchester Section

37 to 29

15.9 km, gravel

Good birding – especially raptors, elephants, zebra, kudu, warthog

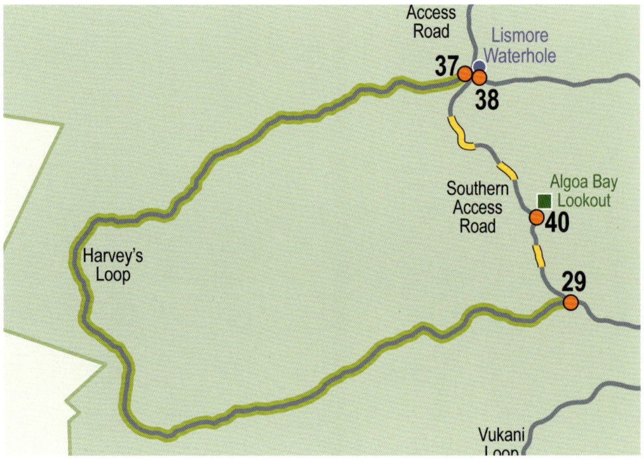

Harvey's Loop is located on the original Harvey farm. Being a gravel road, it can become a bit slippery after heavy rain, but otherwise it is a pleasant drive along healthy thicket vegetation. It follows low-lying areas and wooded kloofs but also circumnavigates open grassland areas where thicket was cleared for farming purposes. The habitat is varied and makes for a good game and birding drive, ensuring it gets a high rating.

A bit of history

When driving along this road you cannot help but think about JT Harvey; who he was and what he was like? After all, he was the man who played a pivotal role in rescuing the last, small elephant herd that had escaped the elephant massacre in 1919 and 1920. He himself was a hunter of elephants, but he was appalled by the mass killings. The last elephants chose the dense thicket on Harvey's farm to hide and he did not expose them. Imagine a world without elephants ... If it hadn't been for him, there would have been no Addo Elephant National Park!

It all started in the early 18th century. Ivory hunters had a heyday collecting tons of valuable ivory and other wildlife products in the Sundays River and Addo area with their guns. By the end of the 19th century hunters had exterminated most of the remaining elephant, lion, buffalo, black rhino and other wildlife. Only small, isolated elephant herds remained in the impermeable thickets at the foot of the Zuurberg Mountains. By 1886, hunters needed a licence to shoot, trap or kill elephant or buffalo, which helped, but numbers were still dwindling. Four years later, portions of Crown Land in the valley and in the Zuurberg were set aside as forest reserves and no more wood cutting or hunting took place.

For a while, farmers tolerated the destruction caused by the remaining elephant and other wildlife. Rhino and lion were already eliminated but elephant and buffalo living in the nearby bush regularly came down to quench their thirst from the sweet waters of the Sundays River, raiding crops and causing havoc at night. The elephants were becoming a problem. In desperation farmers appealed to the government for help and soon a professional hunter was approached.

Major Jan Pretorius was reluctant at first. Addo was known as a 'hunter's hell' and it was believed in hunting circles that because of the impenetrable bush, it would be 'impossible to exterminate the beasts'. Besides, the area was also said to be a death trap as numerous hunters had perished while hunting elephants. Still, Pretorius took up the challenge to rid the area of elephants. From 1919 to 1920, he shot 120 of these giants, often up to five a day. He had a soft spot for elephant calves, which he spared. Eventually only 16 elephants remained. They were hiding in the dense thicket on the Harvey farm. At that stage, Pretorius decided he'd had enough of the carnage and left them be. Harvey felt sorry for these giants and allowed them to stay.

An outcry arose as the public finally realised what was happening and popular opinion turned towards compassion for the elephants. Protests for their protection arose not only in South Africa, but also overseas. When a decision was taken to spare the remainder of the herd, only 11 panic-stricken and bewildered animals were left.

These animals formed the nucleus of the present herd now protected in the Addo Elephant National Park.

There was a positive response to the idea of a national park and in 1931, the park was proclaimed, making it the second-oldest national park in South Africa. Harold Trollope, a former big game hunter and a Kruger National Park ranger, was appointed as the first warden and had the dangerous task of rounding up the remaining elephants and herding them into the new park. Soon the struggle to preserve the Addo elephants was well known both locally and globally. New land was gradually added to give the elephants and other wildlife more space to roam.

The success story is that today the elephant numbers have increased to more than 600. Fortunately, modern veterinary methods allow successful relocation of elephants. This process has already started. Initially 26 elephants were relocated to the Darlington Section and more will gradually follow. Once the elephant-proof fencing of the Kabouga Section is completed, elephants will also be released there. Successful translocations to other wildlife reserves have already been going on for a couple of years.

As a keystone species, elephants play an important role in maintaining the biodiversity of the ecosystem in which they live. Their disappearance will cause permanent, irreversible loss of habitat for many other species.

Red hartebeest (*Alcelaphus buselaphus*)

Stone Age occupation

Harvey's farm is also known for the **Middle Stone Age artefacts** that were found at Addo Heights and Korhaansvlakte. The **Stone Age** marks a period of prehistory in which humans used primitive stone tools. This period ended about 5 000 years ago when humans started working with metal to make tools and weapons. The early history of man in and around the present Addo Park has been researched by the Albany Museum in Makhanda (Grahamstown). Visit the **Ulwazi Interpretive Centre at Main Camp** to see some of the stone artefacts that were found. These are mostly crude with one or two flakes taken out, possibly to sharpen the tool for extracting shellfish off the rocks and get to the meat inside.

Later Stone Age people were ancestral to the San (Bushmen) and Khoikhoi (or Hottentot) peoples. The San were hunter-gatherers who lived at the coast, where they exploited the marine resources such as shellfish, fish, seals and seabirds. Many hundreds of **shell middens** are found along the coast in the park. Inland groups frequently lived in **caves and rock shelters** and there are several sites in the Zuurberg Mountain that testify to this. There are also **rock paintings** in some of these caves. Excavations were carried out at Melkhoutboom and Vygeboom and uncovered graves with rich grave goods indicating a complex belief system. There is evidence that more and more Khoikhoi people arrived and settled in the area between 2 000–1 500 years ago. Many of them died during an outbreak of smallpox that hit the area in the early 1700s.

Game viewing on Harvey's Loop

The climatic conditions in this semi-arid area are extreme. Rainfall is unpredictable and varies between 200–600 mm per annum and can fall any time during the year. Droughts are often followed by floods. Temperature can range from below zero to over 40 degrees Celsius. Local plants and animals must therefore be able to tolerate these extremes to survive. Many of them have special adaptations to rise to the challenge.

Birding is particularly good along this road owing to the diversity of habitats. About 250 bird species occur in the game-drive section of the park and many of these can be spotted on Harvey's Loop. The thorny thicket provides safe roosting and nesting sites. Plants attract birds with their fragrant flowers, nectar, berries and seeds. Others have nutritious leaves.

Expect to see **African hoopoe** feeding on the ground by probing with their long bills in soil while looking for grubs. Another ground-level feeder is the **olive thrush** that obtains prey in leaf debris by scratching and turning over leaves. **Pigeons** and **doves** also feed on the ground by searching for seeds. Look out for **emerald-spotted wood dove**, **Cape turtle dove** and **red-eyed dove**.

Quite a number and variety of birds move around in trees and shrubs foraging for insects and fruit. These include the insect eaters such as the **Karoo scrub robin**, **Karoo prinia**, **southern boubou** and **bokmakierie**. The Karoo scrub robin and Karoo prinia are typical birds of Addo thickets. Southern boubou and bokmakierie normally forage in pairs and both species are famous for their beautiful duet calls. Some insectivorous birds such as **fork-tailed drongo**, **common fiscal** and **flycatchers** hunt from a perch.

Cape starling, white-eyed bulbul and black-capped **bulbul** feed on both animal and plant matter, while the mousebirds are fruit and leaf eaters. The **speckled mousebird** is particularly common, occurs in small swarms and can often be seen sunning itself in the early morning.

On the grassy plains look out for terrestrial birds such as **guinea fowl**, **secretarybird** and **blue crane.** The **black-headed heron** is another common sight and so is the **crowned lapwing**, **ant-eating chat**, **African pipit**, **spotted thick-knee**, **rufous-naped lark** and **Cape wheatear**.

Karoo Scrub Robin (*Cercotrichas coryphoeus*)

Fiscal Flycatcher (*Melaeornis silens*)

Karoo Prinia (*Prinia maculosa*)

Greater Double-collared Sunbird (*Cinnyris afer*)

African Hoopoe (*Upupa africana*)

★★★★

Jack's Picnic Site

Main Game Area

Point: 25 (includes Botanical Reserve)

1 km, gravel

Picnic and recreation area in Botanical Reserve

Jack's Picnic Site gets a high rating because of the superb setting, the excellent birding and plant-spotting opportunities while on foot in pristine thicket vegetation, the clean and well-kept ablutions and safety. In the dry season it gets a bit dusty, but that is to be expected.

The origin of the name

Black rhino have always been part of the scenery in the Addo bush. Unfortunately, they were hunted to extinction by the time the national park was established. Under threat of total extinction of the black rhino in South Africa, black rhino from Kenya were imported into the Addo Park in 1960. It was only established many years later that the black rhino in Africa can be divided into four subspecies and that the Kenyan black rhino was a different subspecies (*Diceros bicornis michaeli*) to the one occurring in the drier parts of South Africa and Namibia (*Diceros bicornis bicornis*). In 1991, the park management decided to stick to the IUCN's decision to stock only subspecies of rhino indigenous to the particular area and therefore the translocation of this subspecies back

Spekboom (*Portulacaria afra*)

Tall Cotyledon (*Cotyledon velutina*)

Jack's Picnic Site

to east Africa's parks started. By this time the population had grown to 46 individuals but by 2003, all the Kenyan subspecies had been removed from the park and reintroduction of desert-adapted black rhinos from Namibia could commence in 1994. The rhino success story was a huge undertaking and is a feather in SANParks' cap.

Jack's Picnic Site plays a part in this story because Jack was one of the original individuals who came from Kenya in 1961, when he was about five years old. His original name was **JA** but he was **renamed Jack** by park ranger Karools Goliath because he did not like the original name that was a reference to JA Hunter, who was responsible for killing over 1 000 black rhino in Kenya. Jack could not be translocated out of the park like the others because of a heart condition. The stress of being sedated and translocated would surely have killed him. The decision was to keep him apart in the fenced Botanical Reserve. Here he spent his last days in peace until he died at the age of well over 30 years; one of only a few black rhino recorded to have reached this age in the wild.

Today, the **Botanical Reserve** is host to many visitors to Addo. The beautiful **picnic site** within this protected plant sanctuary was named in honour of **Jack, the remarkable black rhino**. Here the Subtropical Thicket is at its best since elephants have been kept out for many years and the vegetation is therefore in its pristine state. Each picnic area is a little secluded haven with tables and benches tucked away within the bush, allowing visitors the opportunity to have a private picnic lunch while experiencing the special atmosphere so typical of thicket. Well-serviced ablution blocks are a pleasure.

The picnic site offers wonderful birding. See how many species you can tick off while picnicking. The site also gives you the perfect opportunity to do some plant spotting.

Southern Boubou (*Laniarius ferrugineus*)

★★★★★

Link Road

Main Game Area

6 to 7

1.3 km, gravel and tarred on incline

Grassland, thicket along incline

This is a mere Link Road and runs uphill towards Zuurkop. Half the road traverses grassland and the other half, which is a tarred strip, runs through thicket before it meets up with the Zuurkop Road. It is a surprisingly productive road and offers scenic views into the distance.

Fertile grassland

In the months of drought when the grass is dead and mainly karroid vegetation remains, elephants are rarely seen on grasslands. Once it rains, however, they can't resist the fresh new grasses. In fact, about 30% or more of their diet consists of couch grass, which is the dominant grass species in the park. Elephants use their dexterous trunks to grip the grass close to its roots, pull it out and shake off the soil before it goes straight to the mouth.

Expect to see **zebra**, **red hartebeest** and **warthog** on the grassy plain that runs over the Gorah Loop. Look out for **yellow mongoose** and **suricate**. From time to time **spotted hyena** and **black-backed jackal** are seen scavenging or simply crossing the plain. In Addo, hyena act as important predators, especially for the many rodents, hares and the newborn of larger mammals. Hyena wander far and wide in search of prey and to scavenge.

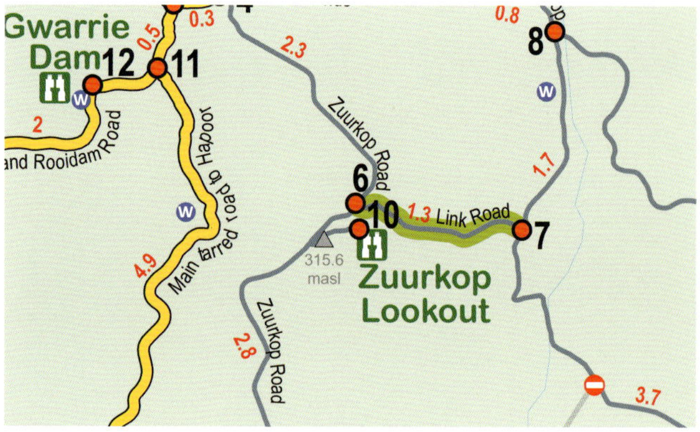

The kings of the Addo castle

Although **lion and hyena** were indigenous in the Sundays River Valley long before the onset of agricultural activities, they were all shot out by the time the park was proclaimed. For many years brown hyena, jackal and caracal as well as the odd leopard were the only predators. Then in **2003, spotted hyena** and later **lion** were brought in to restore the **ecological balance** in the ecosystem. Six of the iconic black-maned lions of the Kalahari were translocated and they adapted well to their new environment.

The main reason for choosing Kalahari lions as opposed to lions from the Kruger National Park, was to protect the **disease-free buffalo** population. The buffalo herd in Addo is the only original disease-free herd in the country. Bovine tuberculosis (TB) is a chronic disease of animals caused by a bacterium *Mycobacterium bovis,* which is closely related to the bacterium that causes human and avian tuberculosis. This disease can affect practically all mammals, causing a general state of illness, coughing and eventual death.

The other reason, is that Kalahari lions are believed to be the closest match to the **now-extinct Cape lion** that used to roam the Addo area.

Blackbacked Jackal (*Canis mesomelas*)

Lion *(Panthera leo)*

Lion numbers must be controlled in a park as small as Addo and interbreeding kept to a minimum to maintain a viable gene pool. If you want to see lion, get up early and be the first to drive into the park. During the day, lion are mainly spotted lazing underneath a bush in the thicket. Locating such a small number of lion during the day is almost like finding a needle in a haystack. A guided-sunset or night drive may also help to spot the lions.

Roads

★★★

Mbabala Loop

Main Game Area

15, 16, 17, 18 to 19

7.3 km, gravel

Kadouw Lookout, Coega Bonteveld

The Mbabala Loop is the flower garden of Zuurkop, but it's not the most productive road as far as game viewing is concerned. Yet, it has a fantastic atmosphere and is quite different from any of the other loops in the park. Do this loop for its tranquillity, wonderful trees, beautiful flowering succulents and bulbous plants. Here you will encounter few vehicles and have the Coega Bonteveld to yourself. Buffalo sightings are reported from time to time, warthog will most certainly be present, but other than that you may see duiker or kudu. There is also a very small chance of sighting a leopard.

Kadouw Lookout

You are allowed to leave your vehicle at your own risk at this lookout. Even so, animals of Addo don't know this. An unexpected surprise awaited a couple a few years ago when they came upon Jack and Jill, the two **lion** siblings, relaxing on the bench enjoying the view. Look around before you alight from your vehicle unless you want to get an unpleasant surprise.

Spend a peaceful half-hour with your binoculars scanning from left to right over the broad valley below. This view is totally different from the Zuurkop Lookout. A century ago the thicket stretched as far as the eye could see into the Sundays River Valley. Now the scene is totally different. Thicket is replaced by agricultural land – mainly citrus – and other industries have taken precedence over conservation.

Although the Kadouw Lookout is not necessarily a favourite with elephants, they do occasionally take a stroll and leave **dung** evidence. This can be an opportunity to study the dung patch for the endemic **flightless dung beetle**. Unlike the rest of the approximate 800 kinds of dung beetles in southern Africa, it has no wings. Instead it has a tightly sealed wing case which creates a convective cooling system where heat is drawn away from its body; and is therefore perfectly suited to the hot climate of Addo.

Originally this beetle was widespread, but today **it is in danger of extinction**. Not only is its habitat under threat of agriculture and other human activity, but **females produce only one offspring** (sometimes two) per **annual** breeding cycle. The beetle feeds on elephant, buffalo and rhino dung and also uses it for breeding. Since the flightless dung beetle of Addo is one of the largest of all dung-beetle species, the **dung ball it makes is quite impressive**. If you come across one of these dung-rolling beetles, watch how it navigates and marvel at its life cycle.

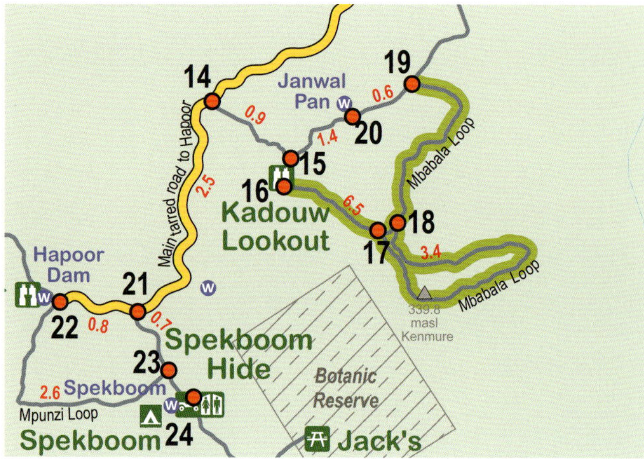

The female works extremely hard and invests a lot of time and energy in the reproduction process, whereas the **male hardly features** during the entire process. Other dung-beetle males are known to roll the dung ball but not this special species. Here it is the **female who constructs the brood ball and rolls it** to a suitable site. Several males will cluster around the ball to attract the female's attention. She is, however, extremely fussy about a mate. It is the female who excavates the hole and when she is ready, the chosen male climbs onto the brood ball and rolls down with the ball into the excavated hole where they mate. Once done, he re-emerges and looks for the next female to charm. She will remain in the hole, lay an egg on a small pinnacle on the brood ball and then start the waiting game. **She remains there for five to six months, without feeding, until the egg hatches** and her babe has reached the young adult stage. While in waiting,

Candelabra Flower

she diligently clears any fungus that might develop on the brood ball. Once the babe is ready to leave the birth chamber, it takes another 50 to 65 days to become sexually mature.

The rare flightless dung beetle is king (or is it queen?) of the road in Addo, with signs warning visitors that this recycling machine has right of way. The beetles are seen only when conditions are not too hot or too cold and play an important role in recycling nutrients; in this way helping the growth of healthy thicket vegetation.

The Loops

'Mbabala' is the local name for **bushbuck**. This is a shy antelope that will seldomly walk out into the open. The male and female look different from each other but both blend in perfectly with the vegetation. It usually keeps close to riverine vegetation with a reliable water source, but here in Addo with the succulent vegetation, it may be able to survive for a longish period without drinking. Bushbuck usually go for fresh succulent shoots and flowers. There is a slim chance of finding bushbuck along this loop.

An unexpected sighting would be a **caracal**. Since they are nocturnal, they are seldomly seen during the day. This is one of the smaller predators that was not eradicated through hunting and was here when the park was proclaimed. Since there are plenty of rodents and hares around, it makes sense that caracal could be seen around this area. They usually stalk their prey but are also capable of leaping up into the air to catch birds. Since they get enough moisture from their prey, they don't need to drink.

Caracal (*Caracel caracel*)

Driving the Mbabala Loop is like doing a meander in a beautiful park. Note the size of the **wild olive**, **white milkwood** and other trees, many of which stand alone here in the Bonteveld. See if you can find crassulas, cotyledons and bulbous plants amid the grass. When in flower, it is easy to find them. Game such as buffalo, eland, kudu, duiker and, of course, bushbuck can be expected here.

Flightless Dung Beetle (*Circellium bacchus*)

★★★★

Mbotyi Loop

Colchester Section

32 to 34

4.6 km, gravel

Antelope, ostrich, Karoo boer-bean

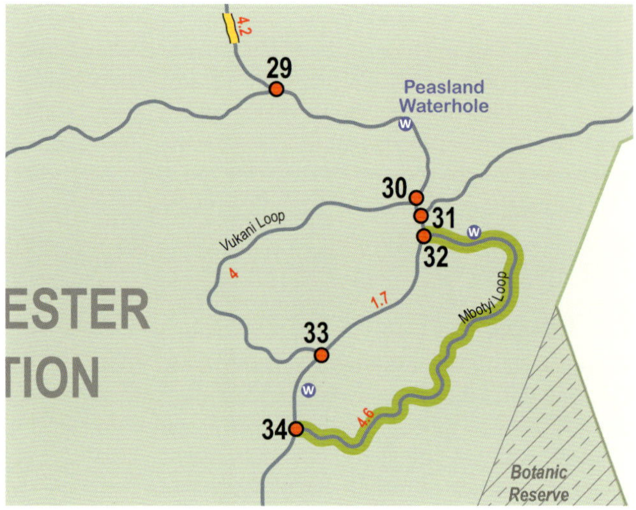

Although Mbotyi Loop is a gravel road, it is scenic with diverse vegetation that offers good game viewing because of open spaces with high visibility. It therefore gets a high rating. At places the thicket is well developed and dense, but towards the top of the low hill, the vegetation opens up into grassland with low scrubby bushes. It is interesting to note the difference in the occurrence of game from the dense thicket to the open habitat. In the dense thicket expect to see the occasional elephant bull, monkeys, the shy bushbuck or kudu, while in the open there may be red hartebeest, ostrich, yellow mongoose and lots of termite mounds. Where there are termite mounds, sightings of aardvark can be expected. The only problem is that they are nocturnal, very skittish and seldom seen during game-drive hours.

Karoo Boer-bean (*Scotia afra*)

Everlasting (*Helichrysum cymosum*)

Vervet monkey (*Cercopithecus pygerythrus*)

The meaning of Mbotyi

A most beautiful gem of the thicket vegetation is the **Karoo Boer-bean** which the locals call '**mbotyi**'. The showy flowers are borne from August to November and are mainly red in colour, but sometimes pink versions are seen. Their nectar attracts both birds and insects, while the seeds are edible and favoured by vervet monkeys.

Addo is not only about elephants

As a keystone species, elephants play an important role in maintaining the biodiversity of the ecosystem in which they live. Yet, if elephant density is too high, this can have a severe negative impact on the structure of the plant community and species richness. One of the park's three botanical reserves is situated to the east of this loop. These reserves have been established to study the impact of elephants and other animals on the thicket vegetation over a long period of time.

Conservation priorities have changed over the years; from the original goal of protecting the last remaining Addo elephants to the protection of all vulnerable species (elephant, buffalo, rhino and the flightless dung beetle) as well as Subtropical Thicket and the protection of overall biodiversity. Therefore, a national park is not about numbers, production and profit; it is about the fauna and the flora in the area. Today the focus of running a national park is based on maintaining a balance of what typically occurs within the area.

Common Ostrich (*Struthio camelus*)

Roads

★★★

Mpunzi Loop

Main Game Area

22 to 23

2.6 km, gravel

Duiker, over-utilised thicket

This short loop starts at the Hapoor Dam and ends just before the Spekboom Hide. Mpunzi is the local name for the common duiker often seen along this loop. The road is not tarred and the first stretch will enable you to see the Hapoor flats from a different viewpoint. The vegetation along this loop shows the effect of the over-utilisation of elephants. It gets an average rating.

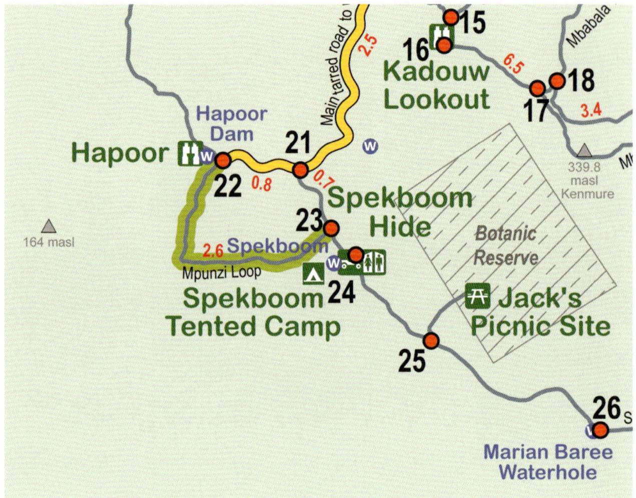

Keystone species

A keystone species is one on which many other species in the ecosystem depend. **Elephants are a keystone species** because they have the ability to change or mould an environment more than any other creature in the park. This can be good or bad. When the number of elephants is in line with the carrying capacity of an ecosystem, they play an important role by opening up dense bush. This makes food available for smaller browsers and stimulates fresh grass growth which can be utilised by other species. Elephants also play an important role in seed dispersal of plants via their dung, which is an important source of food for dung beetles, in particular, the endemic and endangered flightless dung beetle in Addo.

Too many elephants, on the other hand, can alter an ecosystem significantly. Not that many years ago, the thicket along this road and closer to the Hapoor Dam was dense and lush. Now it looks tired and hammered. This is mainly because of over-utilisation of the vegetation by elephants on their way to and from Hapoor Dam.

Water sources

There is no natural river system in the park and therefore **provision of water** in a reserve like Addo is a challenge. The park has to provide and manage water provision by other means. In a natural system you would never find water evenly distributed across the landscape because this would lead to a homogenous landscape and a decline in the diversity of species. In turn, such a scenario will ultimately lead to the extinction or degrading of species. This is what happened at

Suricate/Meerkat (*Suricata suricatta*)

Hapoor. To avoid repeat occurrences around all other water points, management had to make difficult decisions.

Conservation management aims to mimic nature as closely as possible by developing an imaginary water gradient (river). The imaginary river is represented by an imaginary line from Marion Baree Waterhole, Spekboom Waterhole, Hapoor Dam, Rooidam, Gwarrie Dam, Nyati Waterhole (at the Main Camp), Woodlands Waterhole and Domkrag Dam. These water sources supply enough water and dams are big enough to meet the needs of large elephant herds. Elephants are, however, dominant when visiting a water source; chasing away other species, especially the shy ones. Therefore, management aims to reserve the remaining waterholes for the exclusive use of other species and, at the same time, force elephants away from over-utilised areas such as Gorah Loop. This is done artificially through experimental smart fencing of certain waterholes that are not part of the river gradient; to keep elephants out but allowing all smaller game access to the waterhole (examples are Peaseland, Lismore and Carol's Rest Waterholes).

The goal is that elephants will move across the park in the rainy season when all pans and dams have water, but that they will move back to the main water sources in times of drought. In this way the Colchester thicket is protected from being over-browsed. **The number of waterholes in the Colchester Section is therefore purposefully limited as is the quantity of water available**.

In times of plenty, the breeding rate of elephants increases. To reach an equilibrium, elephants need **stress**, **which affects the inter-calving periods**, thus slowing down the growth rate. Droughts assist in causing stress, since water sources gradually dry up. This is absolutely necessary or else the elephant numbers will get out of hand.

Common duiker (*Sylvicapra grimmia*)

Yellow mongoose (*Atilax paludinosus*)

Game viewing

Apart from **elephants** on the way to water, you may see **zebra**, **duiker**, **suricate** and **yellow mongoose**.

If you are early enough, you may spot a group of **suricates** at their warren on the bare area behind the dam. Early mornings they usually warm themselves before leaving for their foraging session. They are social and live in tight-knit groups in which they look after each other, dig their own burrows, sleep together and forage close together. Group members recognise each other by sight and smell. There is usually one individual doing **sentinel duty** and this is more often than not when they are spotted. When foraging they blend into the background and this makes it difficult to see them.

Look out for **the common duiker** that feeds on fresh, juicy leaves and stem tips, flowers and fruit. It has no need for drinking as its food provides all the necessary moisture. Duikers are usually solitary unless it's time to breed. They are territorial and will therefore often be seen in the same vicinity. If you happen to spot a duiker, you will probably see it there again next time you do the drive.

★★★★★

Ngulube Loop

Colchester Section

38 to 31

10.4 km, gravel

Ngulube Waterhole

This is another superb loop in the Colchester Section and has all the elements for earning the highest star rating. The scenic drive leads through valleys between hills and takes you up to the summit of the next incline. Game is abundant, especially in the vicinity of Ngulube Waterhole. Look out for elephant, kudu, caracal, bushbuck, red hartebeest, zebra and warthog. Ngulube is the local name for warthog, of which there are plenty.

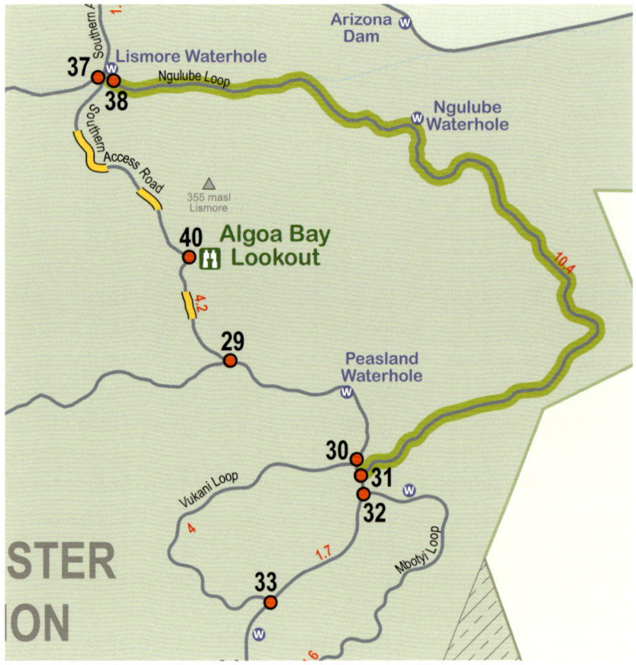

Raptors

Raptors are frequently spotted in the Colchester Section of the park. Some of them are regularly seen along the Ngulube Loop. Raptors are at the top of the feeding pyramid and are the first to suffer when there is environmental degradation. Spotting a raptor is always a special sighting. The raptors mentioned below are still regularly seen in Addo.

Southern pale chanting goshawk is a common resident in the park. They feed on the plentiful Karoo bush rat and other prey items such as insects, reptiles (including snakes) and birds; even as large as guineafowls.

The widespread **black-shouldered kite** hunts either from a perch or by hovering and is well known for its mouse-killing capabilities. Over 90% of its diet consists of rodents, making it one of the most prey-specific raptors. It does, however, also catch insects, reptiles and small birds.

The **jackal buzzard** prefers rocky and mountainous terrain. It hunts from a perch or by stooping in flight and feeds on a variety of prey items such as birds, mammals, insects and reptiles (including venomous snakes), but it also scavenges. This buzzard is often seen along the cliffs of the Ngulube

Common Buzzard (*Buteo vulpinus*)

Rock Kestrel (*Falco rupicolus*)

Loop and in the Zuurberg Section of the park.

Common buzzards are abundant, non-breeding **summer visitors** to Addo. They favour exposed perches, but form flocks at a good food source. They mainly feed on insects, but also catch small birds, frogs and reptiles. Imagine the distance these birds fly from the steppe in the northern hemisphere to a place like Addo, to feed for about three months before returning to their breeding grounds.

Another **summer visitor** to Addo is the **yellow-billed kite**, an infra-African migrant. It feeds on smaller animals and can often be seen scavenging on roadkills. They gather in huge numbers when flying termites emerge after rains.

Black-winged Kite (*Elanus caeruleus*)

Plains Zebra calf and mother *(Equus quagga)*

Plains zebra

The handsome **plains zebra** are frequently seen on the Ngulube Loop. They are boldly striped in black and white and no two individuals look exactly alike. Compared to other zebra species, the plains zebra has broader stripes. They are vertical on the fore part of the body and tend towards the horizontal on the hindquarters. In southern populations, including Addo, zebra have varied but lesser amounts of striping on the underparts, the legs and the hindquarters. Southern populations also have brown 'shadow' stripes between the black and white colouring.

There has been a lot of debate about the purpose of zebra stripes. It is often postulated that the stripes act as camouflage by breaking the animals' outlines, but this does not hold true since they are noisy, active and do not freeze like other animals that rely on camouflage when they are threatened. Others hypothesise that the stripes form a dazzling effect and make it difficult for predators to focus on an individual for the kill, but zebra are successfully predated on by a variety of predators, all with totally different hunting styles. Further studies suggest that stripes may have developed to discourage biting flies and have demonstrated that stripes polarise light in such a way that it discourages these flies in a manner not shown with other coat patterns. Yet, other studies determined that environmental temperature is a strong predictor for zebra striping patterns and proposed that the stripes may be related to thermoregulation. What we do know is the evidence that the **zebra's background colour is dark and the white is an addition**. The first sub-species to be described, the now-extinct quagga, had plain brown hindquarters.

Sparring for dominance

The logo of SANParks was derived from the king of antelopes, the **greater kudu**, with its long and graceful spiral horns that it carries with majestic flair. This antelope is widely admired, not only because of its magnificence and proud posture, but also its agility, acute hearing and eyesight, alertness and above all its adaptability. The kudu truly deserves its iconic status. Male kudus are rarely physically aggressive but may spar during the courtship season, shoving one another with their horns. Occasionally, during these contests their horns become interlocked and, if unable to free themselves, both males may die.

Caracal

Despite its elusive nature, a resident **caracal** is fairly regularly seen along the Ngulube Loop. The thrill of such an unusual sighting is difficult to describe. They prefer mountainous areas, avoid open plains, are nocturnal and secretive. Caracals usually live on their own or in pairs and hunt small mammals, birds and rodents. These beautiful small predators are not a favourite among farmers who often lose livestock as a result of their presence.

Erosion

It is impossible not to notice the **severe erosion scars** along this loop. These scars were formed by the removal of the natural vegetation too close to the drainage line while the runoff from the hill was not controlled before it developed into an erosive force. There is no way these scars will ever be rehabilitated. The scars show the nature of the underlying soil formation.

Greater kudu (*Tragelaphus strepsiceros*)

Heavy erosion is the heritage of former farming activities and over-utilisation

★★★★

Spekboom Road

Main Game Area

- 21, 23, 24, 25, 26, 27, 28
- 9.6 km, gravel and tarred on inclines
- Spekboom Hide, Marion Baree Waterhole
- Marion Baree: S33°31'36" E25°46'38"
  Spekboom Hide & Tented Camp: S33°30'45" E25°45'9"

Spekboom Road is a link between the Main Tarred Road and the Southern Access Road that ends at Addo Heights. Although not tarred, it gets a high rating because it offers good game viewing, is scenic with intermittent good visibility and gives access to the Spekboom Hide (including the tented camp) and Jack's Picnic Site. It also passes the popular Marion Baree Waterhole and Graham's Pan before it gets to the Addo Heights Road and joins up with the Southern Access Road.

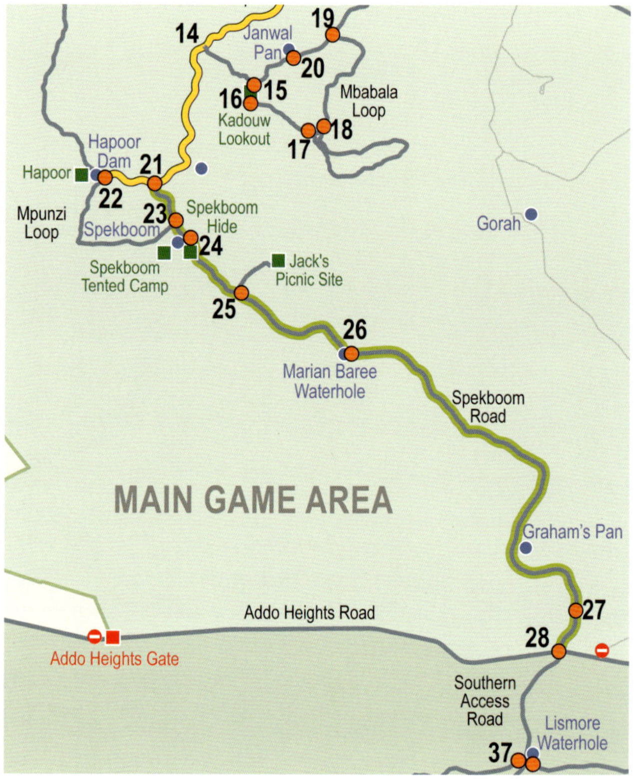

The Spekboom Hide and Tented Camp

It is worthwhile stopping and leaving your vehicle at the **Spekboom Hide**. The entire area is fenced and you will be safe once you are inside the enclosure. Not only will it give you the opportunity to experience the ambience of the thicket vegetation, but you will also be able to watch elephants, buffalo or zebra drinking at the waterhole.

The thicket envelops you as you enter the wonderworld of this strange vegetation type. The silence and a special kind of magic are tangible. Here you may feel what it must be like to be an elephant and why it is so easy for them to vanish once they step into the bush. The **hide is simple but effective** as it allows you to view animals at ground level without them being disturbed by your presence. Take a hat because shade is limited. The hide offers good photographic opportunity and spending time here is worth your while.

Plumbago (*Plumbago auriculata*)

The hide area is extended to the south where the **secluded tented camp** is situated. It is extremely special to spend a night in the middle of the bush, listening to the sounds of the wild surrounding you, catching a glimpse of nocturnal animals and gazing at the stars high above. There are only five tents on wooden platforms, each with two single beds and a trunk with crockery and cutlery. Ablutions are communal. A huge advantage of staying at Spekboom is that once the park gates open, you may leave the enclosure and you are already in the middle of the game-viewing area so you have a better chance of seeing some of the elusive nocturnal animals. Special sightings at the waterhole would be hyena, lion, black-backed jackal, fiery-necked nightjar and spotted thick-knee.

Marion Baree Waterhole

It is easy to drive right past this smallish waterhole which is on the right-hand side (west) of the Spekboom Road. The water is pumped from a nearby borehole and the **inflow of water is controlled**. Good sightings are reported from here. Elephants, mainly bulls and small female herds visit regularly, but also buffalo, warthog and black rhino. The lions that criss-cross the southern areas often use this water source to drink. Good parking is limited.

Look out for lion around here. Although elephants are the main focus when visiting Addo, most people hope to spot a lion or two. Lions tend to select hunting areas where prey is easier to catch rather than areas where prey densities are highest. In Addo, lions are often encountered on the road and early risers have a definite advantage, for once they are disturbed, they simply vanish into the bush.

Southern Masked Weaver (*Ploceus velatus*)

Lion *(Panthera leo)*

Lions originally roamed free across most of South Africa and also here in the Addo region, but hunting and changes in land-use devastated lions in the late 1800s. By 1879, the last lion in the Eastern Cape was shot. By the middle of the 20th century lions in South Africa were restricted to the two largest national parks and even there they were initially regarded as vermin and prosecuted. It was only much later that they became the most valued icons of the bush.

The first **six Kalahari lions** introduced into the park were truly wild. They were unfamiliar with electric fences and did not know each other because they originated from different prides. One of the considerations to bring in lions from the Kalahari rather than from Kruger, is that the Kalahari lions have a valuable disease-free status (from bovine tuberculosis and feline immunodeficiency virus or FIV). When they arrived in 2003, they first spent more than a month in a boma to acclimatise to local conditions and get used to one another and the electric fence that restricted their movements. The lions were fitted with radio collars so researchers could keep track of them once they were released. Of the six lions, two were females and four males. This was done on purpose to minimise the breeding rate.

In a park like Addo, the game area is relatively small and suitable prey is not over-abundant. The lion instinct would drive them to target zebra and antelope, but they will actually take anything from termites, rodents and even baby rhinos. The Kalahari lions were unfamiliar with buffalo and it was hoped they would leave these alone since the disease-free buffalo were a valuable asset that should not be lost as lion food.

A **Predator Management Plan** was therefore established and is strictly followed in Addo. Lions are regularly swapped between parks because this mimics coalition take-overs and improves genetic diversity. The downside to this is that it may lead to infanticide (killing of cubs), which is a natural phenomenon when a new coalition takes over a pride with cubs sired by another male. Contraception is also used to mimic mortality and to control numbers, but is expensive.

Graham's Pan

During rainless periods this pan is dry, but it is interesting that similar **pans dug and deepened by elephants over centuries** exist all over the park. For long periods after sufficient rain, similar pans allow elephants and other game to disperse to areas less utilised during the dry season.

Graham's Pan is situated right next to the road in a valley colloquially known as Wayne's Valley. It is covered by the pioneer **couch grass** favoured by elephants. Grass forms a significant part of the elephant diet. In this vicinity guineafowls are usually encountered. The **helmeted guineafowl** is common and widespread in Africa and the varied vegetation of Addo attracts these birds. They are sociable when not breeding and form large flocks when foraging and resting. During the breeding season males and females separate themselves from the flock to form breeding pairs.

African Buffalo (*Syncerus caffer*)

Roads

★★★★

Vukani Loop

Colchester Section

30 to 33

4 km, gravel

Yellow mongoose, antelope

The highly rated Vukani Loop is a gravel road and another favourite in the Colchester Section. It traverses open areas (former farmland) but also well-vegetated sections. Since visibility is good, game viewing is at its best, especially for grazers. Traffic is low and you will probably have the road to yourself.

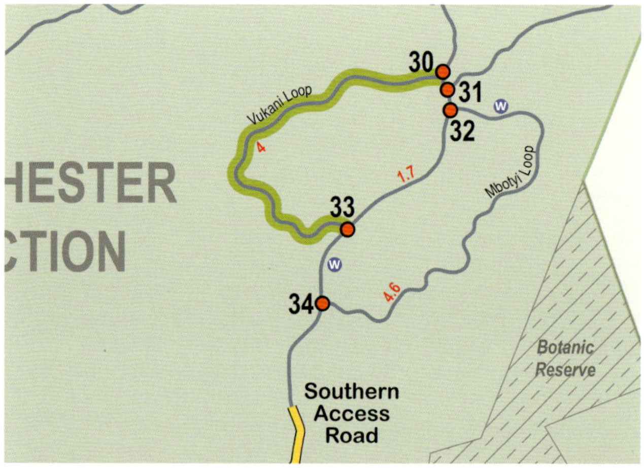

Vukani – the name

Stephen Harold Trollope was the first park manager. Known by the name, **Vukani**, he was responsible for **rounding up the remaining elephants in 1931**. Vukani means 'wake up' in local language. He was an early riser and made sure that his team was up and about the same time he was, so he would shout: "Vukani! Vukani!" to wake them up before sunrise.

The patches of grassland are remnants of the farming activities that took place before the land was incorporated into the park. Once land has been purchased, the process of **rehabilitating the area** begins.

Since 2000, SANParks has received grants from Government (Department of Environmental Affairs and Tourism or DEAT) through a Poverty Relief Programme for the purpose of rehabilitating the newly purchased land. The costs of fencing and improving the additional land that was purchased is often more expensive than the cost of the land itself. The grants received were also used to improve tourism infrastructure, for example, the building of tourist facilities in the Colchester Section.

Yellow mongoose

The Vukani Loop is perfect for watching **yellow mongoose**, which live in social groups ranging from a breeding pair and others of the same family. They sleep and raise their young in dens. One such den is situated in close proximity to the road, a few kilometres from the upper entrance (point 30) on the left underneath a bush. This den is ideal since you can park and just observe their activities. Be there early morning as the mongooses greet each other as they exit the burrow and then sunbathe or groom for a while. The young play around in a rough-and-tumble game, chase each other, mock fight and groom.

Yellow Mongoose youngsters playing

Yellow Mongoose (*Atilax paludinosus*) on Vukani Loop

★★★★★

Woodlands Road

Main Game Area

- 1, 5, 4 to 3b
- 3.1 km, tarred
- Terrestrial birds, eland, brown hyena
- Woodlands Waterhole: S33°26'38" E25°46'16"

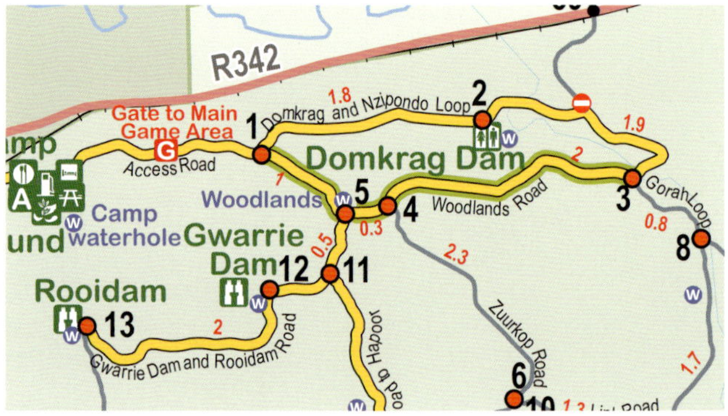

The entire Woodlands Road is excellent for game viewing and can be regarded as Addo's 'Golden Highway'. Although there is some sporadic thicket towards the hills, the wide open areas afford superb visibility. This is the place to find many of the smaller creatures of Addo. Drive slowly and search the veld for scrub hare, bat-eared foxes, springhare (mainly on night drives), yellow mongoose and suricates. You may also encounter lion and buffalo on this road, either at the Woodlands Waterhole or further away among the bushes. Good sightings of shy and seldom-seen animals such as brown hyena, eland and black rhino have been reported. Elephants can be expected anywhere and birding is at its prime. The species diversity is superb and the road condition very good. A well-deserved five-star road.

Eland (*Taurotragus oryx*)

Southern Black Korhaan (*Afrotis afra*)

The birding specials

The **southern black korhaan** used to be a regular sight in the renosterveld of the Western and Eastern Cape but today sightings are rare. This polygynous species, where the **male mates with several females**, is admired for the male's spectacular high-flying display to impress females. One or two dark spotted, olive-brown eggs are laid during spring in a slight depression on bare soil and covered by strands of grass. Only the female partakes in parental care but both male and female will chase away and harass potential enemies that may prey on the eggs or chicks.

In Addo Elephant National Park, the **yellow mongoose** is commonly seen foraging in the same area where a pair of southern black korhaan is regularly seen. All ground-breeding birds despise these small predators and will chase them on sight. Although the yellow mongoose feeds mainly on

various insects, other invertebrates, small rodents, reptiles and amphibians, it will devour eggs or chicks of any ground-breeding bird.

Such a display played out during an extremely dry spell in spring, when grass was absent and only arid karroid vegetation covered the open plain. A visitor was photographing the bird as it leisurely foraged closer and closer to the vehicle. With his finger on the shutter button, he kept the bird in focus while following it in the viewfinder as it moved. Suddenly, the bird changed direction and started charging, running head down while opening its wings in a threatening posture. Only when it started screeching in a loud voice, it became clear what it was after. A terrified yellow mongoose was crouching low on his side, belly up, also growling to warn off the bird. Satisfied that it had sufficiently showed the mongoose whose territory it was, the korhaan regained its composure and carried on while the mongoose walked off in the opposite direction. It was all over in a flash.

Interaction between Yellow Mongoose (*Atilax paludinosus*) and Southern Black Korhaan (*Afrotis afra*)

★★★

Zuurkop Road

Main Game Area

- 4, 6, 19, 15 to 14
- 7.4 km, gravel and tar on incline
- Zuurkop Lookout, Janwal Pan, Coega Bonteveld
- Zuurkop Lookout: S33°27'45" E25°47'5"
  Janwal Pan: S33°29'26" E25°45'51"

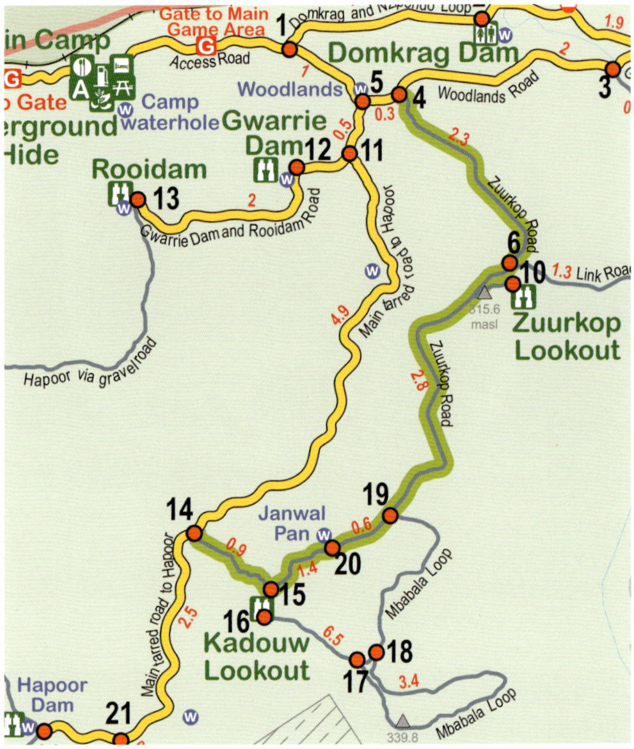

This is a gravel road that takes you to the Zuurkop Hill and Lookout, then leads along the ridge for a couple of kilometres before it descends again to meet up with the Main Tarred Road at the foot of the hill. Game viewing is not particularly good here, but the scenery makes up for that. The road condition is reasonable with some occasional rough patches. This road has an average rating but gives you a good overview of the Addo Elephant National Park and beyond.

The incline towards Zuurkop

The thicket along this slow incline is in good condition and you may see kudu, warthog or even an elephant or two. Try to recognise some of the plants in the thicket community.

Once you reach the top, the road passes a section where a small part of the ridge is washed away. This is a popular spot for a **family of warthog**. They like this open sandy area where they can dust-bathe and there may be a crevice or hole where they sleep. Just beyond this, a pair of **spotted thick-knees** (also known as dikkop) is regularly seen. It is easy to overlook them because they blend so well with their environment. Thick-knees have long yellow legs, speckled plumage and prominent eyes that are also yellow. They occur

Black-footed Cat (*Felis nigripes*)

Spotted Thick-knee (*Burhinus capensis*)

Bushbuck (*Tragelaphus scriptus*)

in this area because the vegetation is open grassland where they hunt exclusively on the ground, feeding on insects, small mammals and lizards.

Zuurkop Lookout and ridge

The **wonderful views** over the northern and northeastern parts of the park make this a particularly scenic road. At the **Zuurkop Lookout** visitors can alight at their own risk and enjoy the views over the green carpet of thicket. The eastern boundary beyond the Gorah Loop is visible on the horizon. In the far distance to the north, the **Zuurberg Mountains**, which forms part of the park, are clearly visible. The undulating landscape below is covered in a blanket of blue-green thicket stretching as far as the eye can see, but also showing the man-made scars that haven't rehabilitated despite the many years since the establishment of the park. Towards the west is the fertile Sundays River Valley with its farming activities.

Imagine what it must have been like for the first 18th century pioneers who occupied this area. Getting through the impenetrable thicket must have been a nightmare. Bush clearing was the only option for building roads, cultivating grazing for livestock and planting crops. It must have been a huge task. Despite the green image, this land is hard to tame with severe droughts, ravaging floods and thorny vegetation posing its challenges. Yet, it is perfect for the plants that grow and animals that live here.

During years with good rainfall, **Janwal Pan** is the only source of water in this area. Expect to see some of the water-independent game along this road, especially in times of drought when the pans are dry. Warthog, kudu and duiker are water-independent game, obtaining sufficient moisture from their diet. They will, however, drink when water is available.

Leopard are very seldom seen in the park but a confirmed sighting has been reported at Janwal Pan. They are shy and secretive animals that easily vanish in the dense thicket. Lately, they are more often seen in the Nyathi Section. This section is, however, not accessible to tourists unless they are booked into the Nyathi Lodge.

Notice that the vegetation along the ridge differs somewhat from the thickets along the first section of the road. The soil on the ridge is poor, shallow and underlain by limestone. This is typical of the unique ecosystem called **Coega Bonteveld**. You will find a fenced-off area, which is an **exclusion plot** to monitor the impact of large herbivores on this type of vegetation. Bonteveld is located on soils of a generally poor quality, occurs in regions with round-the-year rainfall such as parts of the Eastern Cape and is often restricted to 'islands' in a matrix of typical Subtropical Thicket. This vegetation type has been reduced by 90% and remains under severe threat. The species present are a mix of fynbos, grassland and succulent Karoo elements.

One special inhabitant of Bonteveld is on the brink of extinction. This is the tiny **Albany adder**, an easy-to-miss species of viper, only slightly longer than an iPhone. Owing to its beautiful camouflage in tan and brown, it is easily overlooked. The snakes are extremely rare and only 10 specimens have ever been officially documented. They are therefore critically endangered and unless its habitat is protected, the species will disappear for good.

Game such as buffalo, eland, kudu, bushbuck and duiker can be expected here.

Helmeted Guinea Fowls (*Numida meleagris*) protecting chicks from Pale Chanting Goshawk (*Melierax canorus*) attack

Other sections

Other sections

Main Camp Area

Directions: See page 12–13

S33°26'39" E25°44'45"

Tourist area, Ulwazi Interpretive Centre, Ndlovu Waterhole, Birding, PPC Discovery Trail

Nature experiences at Main Camp

The entrance to Main Camp invites you to a hub of activity between game drives. There is something to do for everyone. Some just want to relax, others do birding and some guests visit the activity centre and the underground viewing hide in front of the Camp Waterhole. There is enough opportunity for strolling, hiking, running and visiting the Discovery Trail. Even the youngsters will find camp activities to stimulate their natural inquisitiveness.

First things first

Main Camp Reception is the obvious place to get information and make bookings, but right next to it you will find billboards that show **recent sightings** and other info you may need. At the **shop** you can buy everything from food to mementoes and books on the natural world. The **dining room**, where you can have formal or less formal meals and beverages, is right next to the shop. Opposite reception, a rustic wooden hut houses the main reception area for **guided game drives**. Left of the entrance to the tented camp and caravanning area is a **lapa** that is used for meetings or social occasions. There is a **picnic site** for day visitors and a **swimming pool** reserved for overnight guests only.

Displays and information

Arguably the best place for inquisitive minds is the **Ulwazi Interpretive Centre** situated south of the lapa. This is the place where you can get more answers to your questions. The well-designed and informative display boards offer hoards of information that can lead to a better understanding of conservation, the history of the park, the cultures of earlier inhabitants, the geology and palaeontology as well as special info on the flora and fauna. The trained staff will assist with any more questions.

Outside a **display of skulls and a skeleton of an elephant** helps to put the different sizes of animals in perspective. For the youngsters there are opportunities for a fossil dig, a geology puzzle, whisper dishes, a sandbox and animal tracks; all to raise awareness and educate about the park's biodiversity. During holidays there are usually children's programmes offered at the park.

Ulwazi Interpretive Centre

Camp Waterhole

Camp Waterhole

You can even watch game at the Camp Waterhole, either from a lookout platform close to the picnic area, or from the underground hide. The waterhole is floodlit at night, allowing visitors to spend some quality time just watching and waiting for creatures of the night. Elephant sightings are guaranteed, but predators such as spotted hyena, brown hyena and perhaps even caracal may also be on the cards. Leopards are very skittish and seldom seen in the main game-viewing area, but you may spot genet and porcupine on the outskirts of the circle of light. Buffalo also tend to drink at night and so do black rhino. Quiet evenings are disrupted with the calls of lion, spotted hyena and jackal, while francolin and bokmakierie herald each dawn.

During the day, game viewing depends a lot on the climate and season. When the wind blows elephants tend to hide in the bush and the same applies to cold and rainy weather. When it's dry, all 600+ elephants can be seen dispersed among the waterholes where they drink, bathe and roll in the mud. Drinking antelope are not a common sight in Addo because some of these, like the kudu and bushbuck, get most of their moisture requirements from their food. Zebra drink every day and so do buffalo.

Birding

About 250 bird species occur in the Main Game Area and Colchester Section of the park. Many of these also occur in the Main Camp and can be viewed on foot. The **SASOL Red Bishop Bird Hide** behind the dining room allows views over a marshy area. During the dry season birds are scarce here, but when the marsh has developed again after rain, all kinds can be spotted. The mature fruiting and other trees attract an abundance of birdlife and insects.

In and around the camp, you'll find **Karoo scrub robin** and **Cape robin-chat**, **bokmakierie**, **southern tchagra**, **bar-throated apalis** and **Cape buntings**; also **brown-hooded kingfisher**, **fiscal flycatcher**, **fork-tailed drongo**, **spectacled weaver**, **malachite** and **greater double-collared sunbird** are easily found.

Two related bird species, the **African hoopoe** and the **green wood-hoopoe**, are often seen in the Addo camps. They both have long, curved bills to probe for grubs and smaller vertebrates and both rely on their colours to protect themselves. In nature black, white, green and red colours serve as a warning: "Don't touch me, I am either foul-smelling or bad-tasting."

The **African hoopoe** is a terrestrial forager that walks around in a jerky fashion while probing for grubs in the soil. They nest in hollows of trees or cavities in other objects and since the nest is not cleaned, it becomes foul smelling after a while.

Green wood-hoopoes are mainly arboreal feeders that probe with their conspicuous red bills for grubs and small vertebrates in crevices and behind bark. They are gregarious birds with an alpha female and male that breed in hollows of trees. In the Main Camp, a hole in the old cycad stem in front of the filling station is a favourite nesting site.

Green Wood-hoopoe (*Phoeniculus purpureus*) nesting in dead Cycad stem in Main Camp

Green Wood-hoopoe (*Phoeniculus purpureus*) feeding young

Plant spotting

If you have ever wondered what it must feel like to be an elephant or a kudu in the Addo bush; or perhaps how daunting the bush must have been for the early hunters, a visit to the **PPC Discovery Trail** is a must.

The trail meanders through a section of typical Subtropical Thicket and consists of a boardwalk built from environmentally friendly poly-wood. It is ideal for plant spotting and birding, is accessible to all, including the visually and mobility impaired. The first 500 m of the trail is specially designed to enable easy access for guests in wheelchairs. There are several loops and you can choose how far you want to go. Each loop has an escape route back to the last part of the first loop.

The trail was constructed to blend in with its surroundings and does not detract from the natural beauty of the area. Rest stops along the way feature a bench and information boards providing details on the animals and plants in the vicinity.

As you walk the trail, think and ponder about the uniqueness of this kind of vegetation. Feel the silence and the comforting ambience and bond with the natural world. Try to remember a few names of plants so that you can learn to recognise them in the field. Look for the num-num, the mother-in-law's tongue, the red pelargonium, the firethorn, the common guarri, the melktou, thorny taaibos, common spike-thorn, pig's ear, bear claw, kerky or money plant, sosatie plant, vygies, lichens and old man's beard.

Plant spotting is a wonderful experience, even if you can only recognise five different plants. Imagine being able to identify all the plants in the Addo bush! What an achievement that would be.

Addo's temperature varies between 7°C and 40°C and the unpredictable rainfall is less than 445 mm per year. Plants must be able to withstand such a wide temperature variation and irregular rainfall. Notice therefore, how well most of these plants are adapted to such conditions; many have small leaves or the leaves are totally absent and in some cases, the leaves are leathery and unpalatable. Water is stored either in the leaves, stems or roots and thorns and/or poisonous bitter sap offer protection from browsers. A lot of them can tolerate disturbances from trampling and they re-root quickly or sprout immediately after rain and several have underground storage organs such as corms, bulbs and rhizomes.

Within these adverse environmental conditions the thicket plays a crucial role by acting as a support system to maintain life. The soil in this area is very fine and the thicket is essential to hold the soil and prevent erosion. You may have seen the many lichens, these provide a ground cover and help to break the energy of raindrops. This in turn allows the water to penetrate the soil rather than simply flow away.

You may be lucky to see the spekboom in bloom, but for most of the thicket plants, flowering depends on rainfall. The dominating plants are shrubs between 3–4 m high, while there are few grasses and large trees. Yet, during game drives you will notice that the few tall plants are aloes and euphorbias and that the understorage consists of dwarf succulents and bulbous plants.

PPC Discovery Trail

Black Storm/Desert Spray (*Cadaba aphylla*)

Karoo Boer-bean (*Schotia afra*)

Spekboom (*Portulacaria afra*)

Porcupine (*Hystrix africae-australis*)

Small-spotted Genet (*Genetta genetta*)

Nyathi Rest Camp

Distance from Main Camp: 13 km

Directions: Drive to the Main Camp Reception of AENP to register and get your papers.
Drive from the Main Entrance Gate and turn right on the R335 towards Paterson. Look out for the gate on your left, which is clearly signposted.

Game watching at Nyathi Rest Camp

One of the best places to come across leopard in the Addo Elephant National Park is if you book in at the Nyathi Rest Camp. The self-drive section leads through the concession area of River Bend Lodge and only residents of the Nyathi Rest Camp are allowed to use the 6 km access road that leads to the **most stunning camp** in the park.

The access road traverses a broad **floodplain** at the foot of the Zuurberg Mountain Range with plenty of sweet-thorn acacia and scattered, huge wild fig trees. You are likely to come across elephant, zebra, baboon, vervet monkey and, if you are lucky, get a glimpse of the resident lions and buffalo. Recently, leopard cubs were seen along this access track, which means that such a sighting is not impossible. Note that you are not allowed to leave the access road from the gate to the lodge as this a concession area. The road is not tarred and can get slippery when wet. Along the way you will notice a partly exposed slope that shows the components of the folding nature of the mountains.

The location of this extraordinary **four-star rest camp** is captivating. The individual units are situated on a hill overlooking another broad open floodplain. They are arranged along the crest of the hill so that all units have a lovely view overlooking the grassy plain with a waterhole in the distance.

It is a huge privilege to stay here and guests report fantastic sightings. Imagine solitude, peace, birdsong, comfortable deck chairs to relax and watch game come and go at the waterhole; topped off with a private plunge pool to cool off when the heat gets too much. Towards evening you can enjoy your own braai and sundowners while the sky turns golden and the night activities set in across the valley. **Once there, you won't want to leave.**

Nyathi Rest Camp

The folding nature of the Zuurberg Mountains

A second waterhole is situated next to the entrance gate to the camp. This is the best place to watch **waterbirds**. A boardwalk takes you closer to the water and you can sit and relax there while watching weavers, herons, ducks and geese. You can expect to see several grassland and thicket bird species from your own chalet deck.

A stay at the Nyathi Rest Camp is the ultimate experience in the Addo Elephant National Park.

The Nyathi elephant story

Two of the oldest **forestry conservation areas** could be linked through the purchase of 12 farms north of the Paterson Road up to the Zuurberg Mountain Range. This was called the Nyathi Section and enlarged the park by 14 000 ha. The elephant population in the Main Game Area was increasing and it became imperative to think about transferring some of the elephants from the Addo Main Game Area to the River Bend Concession Area and the Nyathi Section.

This was easier said than done, but the conservation team came up with a good plan. In April 2003, the road between Paterson and Addo (R346) was temporarily closed and the railway line shut down. A corridor was built over the road from the one section to the other and fenced off with a temporary Armstrong fence. Meanwhile, a group of related elephants with their matriarch Anvil had been isolated in the Kleinvlak area of the park over a period of several months. Once the corridor was ready, a helicopter and game capture vehicles were used to drive the elephants towards the corridor. Most allowed themselves to be herded, but 11 refused point blank to be driven over the road. The only solution was to dart them so they could be moved by game capture vehicles to the other side. They settled well, but a new infusion of genes was necessary to prevent further inbreeding.

In 2006, four elephant bulls from the Kruger National Park were translocated and released into the area. Buffalo, red hartebeest, ostrich, eland, zebra and lion were also introduced into the Nyathi Section. Baboon, vervet monkey, caracal, leopard and brown hyena occur naturally.

Lesser Striped Swallow (*Cecropis abyssinica*)

The Zuurberg Section and Narina Bush Camp

Zuurberg Section

±30 km; Time: ±40 minutes

S33°21'4" E25°44'33"

Directions: Take any of the three routes to the Main Camp, register and get a permit at reception. It is possible to get a permit at the Zuurberg Section Office, but they have no card facilities and handle cash only. The road is gravel and even though it is not necessary to have a 4x4 to use it, a high ground-clearance vehicle is needed.
Exit the main gate and turn left.
±1 km down the road, take the turn-off to the Zuurberg Mountain Inn.
Follow the gravel road (Zuurberg Pass) for ±15 km. Take care – the road is narrow and winding.
On your right you will see the Zuurberg Inn/Hotel, but the turn-off to the Zuurberg Section is to the left.
Follow the narrow gravel road until you reach the park offices where you can park your vehicle.

Crowned Hornbill (*Lophoceros alboterminatus*)

Game viewing and plant spotting at Zuurberg

The self-drive route to the Zuurberg Section of the park ends at the **Zuurberg Office**. From there you will have to investigate the area **on foot** or on **horseback**. For hikes, a conservation fee is payable. Horseback rides must be booked at Main Camp Reception. There is ample and secure parking for your vehicle.

Even if you are not a hiker or a horse rider, it is still worth the drive. The route leading to Zuurberg is a public dirt road (R335) and gets narrow at places where it meanders up along the Zuurberg Pass. The views of hill upon rolling hill of thicket-clad slopes and valleys are beautiful to behold. The altitude increases from the foot slopes at 150 m AMSL (above main sea level) to the summit which only reaches 951 m AMSL. The Zuurberg Range is therefore not particularly high, but is sufficient to form a rain shadow to the north.

The east-west trending mountain range is part of the **Cape Fold Belt**, and dominates the northern vistas of the park. The geological formation is clearly visible at different locations in the park. Some of these folds can be seen in areas where the rock is exposed, especially in the Kabouga Section. The mountains consist of hard quartzitic rock of the Witteberg Group with narrow bands of shale. The soils are generally more fertile and finely textured than soils of the Cape Fold Belt to the west.

Notice how the vegetation along the road gradually changes as you drive towards the summit. Before you get to the pass, you will see a healthy stand of **tree euphorbia** along rock faces. This tree euphorbia species is native to dry areas in southern Africa. Its milky latex is poisonous and a dangerous irritant. Once you reach the top of the mountain, fynbos vegetation becomes dominant. The mean annual rainfall is generally in excess of 600 mm and peaks in spring and autumn. During summer thunderstorms occur and lightning often causes fire. This is unlike other fynbos areas further south.

Crowned Hornbill (*Lophoceros alboterminatus*)

Pincussion Protea (*Leucadendron sp.*)

Pincussion Protea (*Leucadendron sp.*)

Hottentot's Tea (*Helichrysum sp.*)

Crossberry (*Grevia occidentalis*)

Peeling Plane (*Ochna pulcra*)

Senecio (*Senecio rosmarinifolius*)

Cape Honeysuckle (*Tecoma capensis*)

Sweetthorn Acasia (*Vachellia karroo*)

Curved Aloe in the Zuurberg

The Zuurberg Pass and a bit of history

The Zuurberg Pass was built between 1848 and 1858 by hundreds of convicts. Several outspans (free resting places for travellers and their wagons) operated in the area at the time. One of the earliest was Hubbard's on the banks of the Sundays River at Addo Drift. Another was situated at what is now Coerney Station, while Stubb's and Boontjies River were outspans on the Zuurberg Pass. Matthew's Inn is today the **Zuurberg Mountain Village** and the erstwhile Webster's Inn was at the foot of the northern slopes where **Ann's Villa** is today.

The entire scenic R335 can be driven up to Ann's Villa and beyond, but after the Zuurberg Mountain Village, the road becomes a bit rough and, at the time of writing, construction is taking place to eventually tar this original route over the Zuurberg Mountains into the Karoo.

The **Manor House** at the Zuurberg Mountain Village (GPS co-ordinates: S33°21'071" E25°44'674") dates back to **1861**. Both this lovely hotel and the Zuurberg Pass could tell us many stories of yesteryear if only they could speak. One of them was the trigger for the **Fourth Frontier War** between the British and the Xhosa.

The Xhosa Wars were a series of nine wars or flare-ups between the Xhosa Kingdom and European settlers in what is now the Eastern Cape. These events were the longest-running military actions in the history of African colonialism. Following the re-establishment of British authority at the Cape in 1806, the British found themselves increasingly drawn into the conflict between the indigenous peoples and the settlers. By 1810, the major confrontation was with the Xhosa on the eastern frontier. During December 1811, Landdrost **Anders Stockenström** and a party of 40 men were stationed near Ann's Villa to prevent the influx of indigenous people towards the Graaff-Reinet interior.

On **28 December 1811**, Stockenström and his men came across a group of Xhosa and a few Khoi near the peak of the Zuurberg. Apparently Stockenström intended to negotiate with the Xhosa to vacate the Eastern Cape area and move back to beyond the Fish River in exchange for a sum of money rather than fighting another war. Instead, a battle ensued and Stockenström and his men were massacred at **Doringnek** with assegaais (spears). All were killed except two, who managed to escape unnoticed, but wounded.

Stockenström was mistaken to be the head of the British Army and his attackers thought that by killing him they would collapse the British war effort. This act led to the start of the **Fourth Border War of 1811 to 1812** in which 20 000 Xhosa under Ndlambe were forced back across the Fish River by the British Army under the leadership of **Colonel John Graham** assisted by Piet Retief and his commando from Stellenbosch. Piet Retief later married the widow of Anders Stockenström. A **memorial plaque** to commemorate this event was recently erected at Doringnek.

The Zuurberg hikes

Starting from the Zuurberg Office, there are two hikes to choose from. The short **three-kilometre**, **one-hour** hike is called the **Cycad Trail**. This is a short circular trail that takes you along a cycad-clad slope down into a valley. As you pass the first cycads (these are the endemic **long-leaved cycads**), marvel at these 'living fossils', which have a long geological history. Once you reach the forest margin, look out for **crowned hornbills** often seen here. Entering and hiking in the **Afro-montane forest** and along the clear mountain stream is heavenly. These forests house the biggest concentration of blue duiker in the park.

The **Zuurberg Trail** is the longer **11 km** hike that takes about **three hours** to complete. This is a challenging walk, but the scenery is breathtaking and well worth the effort. A pleasant surprise awaits once you reach the **Blougat Pool**. This is the perfect place for a picnic and a dip into the mountain pool. From here, there is a shortcut back to the Permit Office.

If you are hiking fit and wish to do the slightly longer **Doringnek Trail** (named after the Doringnek Massacre), you have to carry on further until you reach **Otto's Pool** before you start the return part of the trail. The return climb is taxing, but a sighting of a **red hartebeest herd** makes it easier. Once back on the plateau, look out for the magnificent king proteas and other fynbos species. This trail covers **12 km** and will take about **five hours** to complete.

Imagine walking the hills where ±1 100 plant species have been recorded and **four of the five biomes** of the park are represented. **Thicket** occurs on the lower hills but the hike crosses **grassland**, **fynbos** and **Afro-montane forest**. Recorded endangered species include three species of cycad, two yellowwood species and the rare and endemic **Zuurberg hunchback** or donkey ears.

The Zuurberg Mountains support **significant numbers of game**, including mountain reedbuck, red hartebeest, eland and kudu, as well as Cape mountain zebra, bushpig, baboon, vervet monkey, dassie, and black-backed jackal. You may also encounter the shy blue duiker, the tiny African hedgehog, a frolicking clawless otter in the pools and along the streams, aardvark diggings at termite mounds and sometimes even a honey badger. Look out for African crowned eagle, forest weavers, Knysna turacos and Narina trogons. In summer, you may find emerald cuckoos and blue-mantled crested-flycatchers.

The combination of clean air, pure silence, breathtaking views and healthy exercise will leave you fully revived.

Zuurberg landscape

King Protea (*Protea cynaroides*)

Narina Trogon (*Apaloderma narina*)

Narina Bush Camp

±27 km; Time: ±45 minutes

S33°21'5" E25°41'22"

Directions:

- Take any of the three routes to the Main Camp, register and get a permit at reception. The road is gravel and even though it is not necessary to have a 4x4 to use it, a high ground-clearance vehicle is needed.
- Guests should arrive at least two hours before sunset as there is no electricity at the camp. Bring your own torch/ flashlight. Bookings are made through Central Reservations.
- Exit the main gate and turn left
- ±1 km down the road, take the turn-off to the Zuurberg Mountain Inn
- Take the first turn to the left signposted Enon/Slagboom
- Follow this road and then turn to the right where it is signposted Slagboom
- Continue straight and pass the turn-off to Slagboom Police Training Camp
- Keep following this road, passing through a number of gates, until you see the turn-off to Narina Bush Camp on your right. Leave your vehicle at this parking lot and walk ±500 m to the camp across the bridge over the river and through the forest.

Staying at Narina Bush Camp

Access to the **Narina Bush Camp** is not along the same road as the access to the Zuurberg Office. It is situated at the **foothills** of the Zuurberg Mountains and on the banks of the **Wit River**, a tributary of the Sundays River. None of the large wildlife species, such as the Big Five, occurs here and the area is quite safe as far as wild animals are concerned. Narina offers a **tranquil getaway in a magical forest**, rich in birdlife and smaller mammal species such as tree dassie, blue duiker, bushbuck and bushpig.

Apart from relaxing and swimming, hiking and horse riding are the most popular activities. The hiking trail meets up with the Zuurberg Trail which starts at the Zuurberg Office. Bookings for horse rides are made at Main Camp.

Self-drive the Kabouga Section and Bedrogfontein 4x4

Distance from Main Camp: ±53 km; Time: ±1 hour

The Kabouga Gate: S33°22'39" E25°27'33"

Directions:

- Exit main gate and turn left
- Follow the R335 towards the town of Addo
- Before entering the town, turn off to the right into the R336 towards Kirkwood
- Follow this road until you reach the town of Kirkwood
- At the four-way intersection continue straight and travel over a small bridge
- Turn right after the bridge at the T-junction and follow the road as it becomes gravel, passes a township and graveyard, until you see the signage to the Kabouga Office to the right.

The **Kabouga Section** of the Addo Elephant National Park is adjacent to the Zuurberg Section and also stretches across the Zuurberg Mountain Range. This is perhaps the least known section of the park, although it has been protected since **1896**; first as a forest reserve and later as a national park **(1985)** before it was eventually integrated into the Addo Elephant National Park **(1994)**. This is the place to go if you want **solitude**, **adventure**, **scenic beauty** and **tranquillity**.

The original **Zuurberg Forest Reserve** was proclaimed in 1869 with the intention of protecting a characteristic sample of the grassy fynbos and Afro-montane forest patches distinctive of the Zuurberg Mountains. Woodcutters were decimating the indigenous forests and hunters were destroying the game. At the same time, more and more people were moving to the town of Gqeberha (Port Elizabeth) and the demand for timber and firewood was high, especially for the firing of kilns for brick-making. Woodcutters were turning to the Zuurberg Forests for their needs. Devastation of forested areas would affect water catchment because the Sundays River is fed by the Kabouga, Krom, Groot Ui, Klein Ui and Wit Rivers, which arise in the Zuurberg Mountains. Realising all this, the Department of Forestry decided it was time to protect some of the magnificent forests in the Zuurberg.

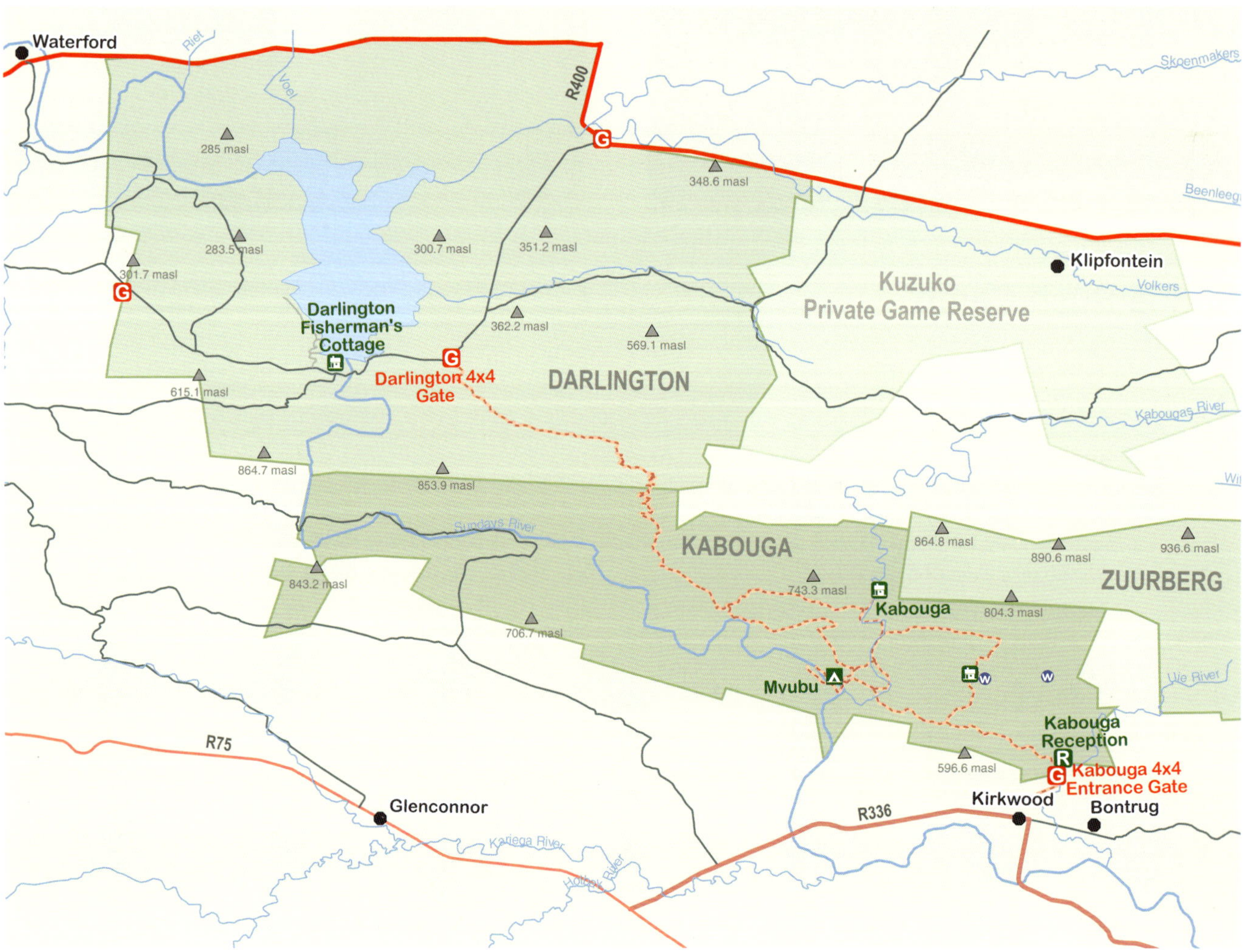

Common Ostrich (*Struthio camelus*)

Yellow-green Cotyledon (*Cotyledon campanulata*)

Kabouga Cottage

Self-drive in the Kabouga

Although the Kabouga has a long conservation history, it is still in the developmental phase for regular tourism and self-drive game viewing. At the time of writing game fences were still being erected and with time, more of the surplus animals from the Main Game Area will be relocated here. Self-drive routes are gradually being upgraded and more tourism infrastructure is being planned. The **Kabouga Office** is the starting point for the **Bedrogfontein 4x4 route** over the mountain, but it is also the entrance to the self-drive routes that are in a developmental stage.

Overnight accommodation is presently restricted to the **Kabouga Cottage** and the **Mvubu Camping Site**. The network of roads on the lower foothills and valleys can be accessed by a high clearance vehicle (not necessarily a four-wheel drive). The upgraded Bedrogfontein 4x4 is now a **one-way trail**, and starts only at Kabouga Gate. Get your permit at the Main Camp Office or pay at the Kabouga Office.

From afar, the Zuurberg Range is a blue haze against the horizon, but as soon as you enter the gate, the romance starts. The feeling as you follow the road along the first valley, is incredible. Like entering the heart of a magical place, steep, awe-inspiring hills enfold you from both sides. The **Valley Thicket** at the valley basin includes all kinds of huge fully grown trees such as sweet-thorn acacia, African olive, karee, cabbage trees, wild fig trees, white-stinkwood, jacket-plum, white-milkwood and others. The **drier northern slopes** are spectacular in the winter when the huge single-trunked **tilt-head aloes** are in flower. If you are a regular visitor to AENP and never been enveloped by these mountains, it is time to go and investigate.

After about 4.2 km you will pass the turn-off to the ranger's house. Another 2 km further the road splits towards the Woodlands Loop and is a 7 km detour.

On the way to the Mvubu Campsite, look out for ostrich, warthog, baboon, red hartebeest and kudu. The campsite is situated on the banks of the Sundays River and there is a possibility of finding a pod of **hippo** there. They were introduced in 1992. Camping in this place is a spiritual experience. It is a totally secluded, rustic campsite with only the river and wildlife. Fishing is allowed. You may even see the shy termite-eating aardwolf, blue duiker and other nocturnal creatures.

The **African crowned eagle** breeds in the wooded kloofs of the Zuurberg. Scout for typical forest species such as olive bush shrike, yellow-throated woodland warbler and Cape batis.

The **Kabouga Guest House** sleeps maximum five and is situated in a secluded valley surrounded by magnificent green hills. Only overnight visitors are allowed up to the house, which overlooks an exposed rock face that clearly shows the geological nature of the fold mountains. Use your binoculars to watch the hyraxes (also called dassies) clamber up and down to find a place in the sun. Parts of the old garden are still visible with age-old cycads leading the way to a natural spring where a moss-clad farm dam was built in yesteryear to collect water for irrigation. Surrounded by yellow-wood, gnarled wild fig, shady wild olive, karee and others; with spring water flowing into the dam, the magic of a forested kloof becomes tangible. It's the ideal place to relax, have a braai and a swim together with the mountain frogs and other water creatures.

Two hiking routes into the nearby hills start at the guest house. The **Kabouga Hike** is only about 400 m and leads to a wonderful lookout point over the valley. The other is slightly longer and is called the **Blue Duiker Circular Route**.

Kabouga and Darlington will soon become a Big Five area once the separating fences are dropped. This will bring about several changes.

Stone Age people

Stone Age people inhabited the park in the **Middle Stone Age** (MSA) between 125 000 and 30 000 years ago. Until about 20 000 years ago there was presumably little human habitation in the area, but during the subsequent **Late Stone Age**, human populations were established in a series of phases.

The **San people were hunter-gatherers** and occupied the area until the Khoikhoi (aka Khoekhoen) people arrived 2 000 years ago. The San people are considered to be direct descendants of the early humans. The **Khoikhoi people were herders** and migrated to the south with their sheep and cattle, then later mixed with the San to form the group of people we call the **Khoisan**.

Scatters of MSA tools are reported along the Sundays River Valley and also inland at Addo Heights and Korhaansvlakte. These sites contain well-preserved plant remains which indicate how they utilised their environment. **Artefacts and graves** dating back 15 000 years have been excavated at places like Melkhoutboom and Vygeboom. This indicates the Cape Fold Mountains was a prime focus of prehistoric settlement. Animal remains found in these caves indicate that steenbok, grysbok and kudu were eaten and that they were hunted by snares.

Greater Kudu (*Tragelaphus strepsiceros*)

Fig Marigold/Vygie/Mesem (*Mesembryanthemaceae family*)

Stone tools were used for hunting, cutting up carcasses, scraping hides, leatherwork and woodwork. These tools are generally very small and indicate they may have been mounted on a bone or wooden handle. By exploring the cave floors, researchers have found signs that the hunter-gatherers occupied these mountains in the summer months when veld food was abundant while they spent the winters exploiting the coastal areas. Many **San paintings** have also been found in various parts of the Zuurberg Mountains.

Numerous place names in the park are derived from the Khoikhoi. The name 'Addo' is thought to derive from the Khoikhoi word '!Ga-dao', pronounced 'gha (with a click) dough', meaning drift (dao) **where the poisonous noorsboom plant (!Ga) grows**. This later became Kadouw and today we talk about Addo. Other place names that confirm the former presence of Khoikhoi are Coerney (originally Koernoe), Nanaga, Boknes, Gorah, Kabouga, Kariega and others.

By the 1820s there were no hunter-gatherers left in the Zuurberg Mountains. Most of them had moved away or were absorbed in the colonial lifestyle, becoming farm workers for the Dutch and British or clients of the Xhosa, where they were engaged in elephant hunting. Most of the Khoikhoi and San were wiped out by a smallpox epidemic and human persecution. Others settled at mission stations such as Enon, Bethelsdorp and Theopolis. For all of them it was an abrupt end to centuries of a hunter-gatherer lifestyle.

Bedrogfontein 4x4 route

The Bedrogfontein 4x4 route

Sir Harry Smith built the original pass that crosses the mountains and is now known as the Bedrogfontein 4x4 Route. The name Bedrogfontein ('deception fountain' in Afrikaans) was derived from a hidden fountain in the kloof that is difficult to find because it appears to be dry but runs underground, leaving no trace on the surface.

Bedrogfontein was also the site of a skirmish during the second Anglo Boer War, when General Jan Smuts and a commando of 250 men passed through the Sundays River Valley on their epic campaign to the northwest. They attempted to ambush a British convoy of 2 000 men transporting cannons to the Free State. None of the Boers died in the battle, but about 700 horses were killed. Apparently a terrible smell of decay in the kloof kept everybody away for weeks on end. There are still some traces of the battle around if you know what to look for.

The Bedrogfontein 4x4 Route is something special and if you have a 4x4 with high clearance, you should not miss it. The scenes are stunning, the vegetation amazing and wildlife abounds. The route itself is not all that difficult and is rated from easy (most of the route) to moderate, with short but difficult stream-crossing sections rated at three out of five (only about 15% of the route). It is not a good idea to drive the route after heavy rain when the streams are swollen, but normally they pose no problem. The route may be driven only from east to west and takes between five and six hours, excluding stops and any side diversions. The 45 km route is suitable for intermediate and experienced drivers, but it is recommended – although not mandatory – to drive in a small group of at least two vehicles in case of a breakdown.

Be prepared and take your time, drive slowly and enjoy the scenery. The route is clearly signposted and easy to follow. While in the foothills, the roads may be slightly confusing if you wish to include the difficult Sundays River Crossing. Before you decide on this, ask about the water level at the gate. This may be quite a difficult crossing and is generally rated as a grade five. All the intersections are well signposted in the form of green concrete triangles close to the roadway.

The rest of the stream crossings along the main route are fairly easy, although the entry and exit angles may be steep. Beware of sharp corners, ruts, sharp stones and water erosion where the road surface is sandy. There are plenty switchbacks, hairpin and horse-shoe bends. Make sure your spare wheel is in working order. When crossing the stony dry riverbed, be careful that the underparts of your vehicle don't connect with the rocks.

There are a few false summits before you reach the real summit at 855m AMSL. Expect some steep ascents and descents with magnificent views into the distance. You will travel through grassland, Subtropical Thicket, beautiful kloofs with Valley Thicket and Afro-montane forests. Nearing the summit of the hills you will find fynbos with proteas and ericas. Take something warm if you want to spend time on the highest summit before your descent into the Darlington Section (it is always cold there). Eventually you will descend into typical Nama Karoo vegetation with shepherd's trees, coral aloes, noorsboom and other succulent plants.

Don't miss the cycads, the living fossils of the Zuurberg Mountains. You will come across two types of cycads growing on the northern slopes of the highest peak. There is the high growing, darker **long-leaved cycad** and the shorter, grey **Karoo cycad**. The Karoo cycad is described as the hardiest, most drought resistant of all cycad species. It is low-growing, small to medium in size, has an erect trunk and forms clumps of up to 10 stems. Old leaf bases form ring patterns on the stem. The leaves are a beautiful blue-grey, 1–1.5m long and appear in whorls. The leaves become entangled as the plant becomes bushy and the crowns appear to lose individual identity.

The long-leaved cycad can grow up to five metres tall with up to 10 stems but is usually a medium-sized clumping cycad. The leaves are dark- to blue-green, numerous, 1–1.8m in length, with leaflets arranged in a distinct V-shape pattern that overlaps towards the tip. The young leaves are covered in fine hair but this is lost as the leaves get older.

Long-leaved/Zuurberg Cycad (*Encephalartos longifolius*)

Propeller Plant (*Crassula perforate*)

Karoo Cycad (*Encephalartos lehmanii*)

Darlington Section

Distance: 134 km; Time: ±2 hours

Directions: Recommended route

- The easiest route to get to the Darlington Section is to leave Main Camp and turn right into R342 in the direction of Paterson.
- At the Paterson intersection with the N10, turn left onto the N10.
- At the intersection with the R400, turn left and continue until the signpost to Addo Elephant National Park, Darlington Section appears. A huge section of this route is tarred.
- Pass through the entrance gate and continue with the provincial gravel road up to the Darlington Dam Wall. There is not much else to justify a self-drive excursion.

Directions: Alternative route

- Do the Bedrogfontein 4x4 Route starting at the Kabouga Section. The last leg of the route passes through the Darlington Section. The easiest way out and back to Main Camp is to turn right when hitting the provincial road and return via R400 and N10 to Paterson before taking the R342 to Main Camp.
- Turn left at the provincial road as you leave the 4x4 route if you wish to return via Kirkwood on the R75.

Chacma Baboon (*Papio ursinus*)

Game at Darlington Section

The Darlington Dam Section of the Addo Elephant National Park is the **largest terrestrial section** of the park at **45 000 ha**, which surrounds the **Darlington Dam** (previously known as Lake Mentz). Self-drive game viewing in this section is mostly limited to tourists returning from the 4x4 Bedrogfontein Route on their way out of the park. A provincial gravel road traverses the park and that is the only road that offers game-viewing opportunities to the public. Tourism facilities are on the cards for the future, but there is still a lot of work to be done before all the proposed infrastructure will be ready. In the meantime, it is good to know about the wonderful conservation actions that are taking place there.

The Addo Elephant National Park fringes the plains of the huge Karoo Basin in the north. Here summer rainfall is often as low as 400 mm a year, of which the typical stunted vegetation of dwarf scrubland interspersed with grasses stands testimony. Despite the semi-arid conditions and extreme temperatures, life flourishes. This region supports the highest diversity of plant species in the Karoo. Overgrazing posed such a major threat to the Nama Karoo biome, that it was imperative to include the Darlington area under the protection of a national park and therefore a large section of this biome was incorporated in the AENP.

Most of the former farming areas surrounding the Darlington Dam have been incorporated into the national park. Large herds of black wildebeest, red hartebeest and springbok roam the floodplains of the dam. Other antelope that enjoy the Nama Karoo vegetation include gemsbok, steenbok, mountain reedbuck and klipspringer. Mountain zebra, baboon, vervet monkey, small carnivores and leopard add to the diversity. The entire **Darlington area is now fenced in** and elephants from the Main Game Area have recently been translocated here. Black rhino do exceptionally well. **Noorsboom** is an euphorbia species that grows profusely together with aloe and stunted shrubs and is relished by the black rhino.

Cape Mountain Zebra (*Equus zebra zebra*)

Darlington Dam

Birding is also exciting with **Karoo endemics** such as pririt batis, rufous-eared warbler and Karoo chat to look out for. The Darlington Dam hosts Goliath heron, African fish eagle, lesser flamingo, terns and grey-headed gulls.

The history of the Darlington Dam begins in **1917** with the decision made by Parliament to build a storage dam, mainly to supply the citrus farmers with irrigation water. Sir Percy Fitzpatrick and the Cape Sundays River Settlements Company (formed in 1914) were instrumental to this decision. The dam was named Lake Mentz in honour of Colonel Mentz, the then Minister of Lands, and construction began in 1918. Lake Mentz was renamed **Darlington Dam in 1995**, after the settlement that was overrun when the dam was created. The dam became part of the Addo Elephant National Park in **2000**.

The following numbers give an idea of the extent to which the Darlington Section has already been stocked with game: Typical **Karoo antelope** that only occur in the Darlington Section include springbok (1 363), Cape mountain zebra (137), black wildebeest (174), gemsbok (142) and red hartebeest (132). A small herd of buffalo (27) has been introduced and so were 28 elephants. The kudu and eland are doing well in most of the sections and in Darlington, the numbers stand at 895 for kudu and 128 for eland. Ostriches (655) are extremely versatile and also occur in most sections of the park.

Translocation of a family of elephants

Elephants that used to roam this part of Addo Elephant Park for eons have been absent for over 150 years. On **31 July 2018, the first family group of 13** set foot here again. The translocation truck arrived, pulled up near the dam and prepared for the off-load. The onlookers were dead silent. The door of the truck was opened, the first cow emerged without a sound, and the rest of the group followed and vanished silently into the night. The next morning there was no sign of them; everybody expected them to be at the dam, but not a trace was seen. They were simply gone as if they'd never been dropped off. Eventually, the group was found in a sweet thorn thicket near the river, north of the area where they were released.

The second group arrived later during the next day and the third and final group arrived the day after. They also vanished soon after they were released.

Karoo Shepherd Tree (*Boscia oleoides*)

Noorsdoring (*Euphorbia radyeri*) in flower

Ten days later, the entire herd was found in the southeastern corner of the Darlington Section where they had discovered thicket. The most extraordinary thing happened while they were watched by the conservation team who found them; the matriarch called out to the herd only once and started walking off, with every single elephant responding to her call and following suit. The team also followed after a while and discovered that she had led them to a nearby waterhole to drink. In December of the same year, the first calf was born in Darlington and during 2019, the second one made its appearance. The herd stayed more or less in the same area for a long time, not venturing too far except going for a drink at the waterhole.

By monitoring their movements since then, it became clear they were walking up and down along the elephant-proof fence, probably trying to find a way to escape back to their former home. Driving a 4x4 down towards Darlington from the summit of the mountain, one can see the path the elephants have trodden to find a way home. Elephants are indeed a super-intelligent species.

Elephants in flat lowlands at Darlington

Black Wildebeest (*Connochaetes gnou*)

Springbok (*Antidorcas marsupialis*)

Woody Cape and Alexandria Forest

105 km; Time: 1 hour, 15 minutes

Directions:

To get to Woody Cape you have to leave Addo Main Camp, or wherever you are departing from, get onto the R72 and drive to the town Alexandria. Look out for the **Addo Elephant National Park signboard** at the western end of the town. A dirt road takes you south towards one of the most beautiful, rare and most threatened forests in South Africa.

- Exit main gate and turn right onto the R342.
- Follow R342 to the Paterson intersection with the N10.
- Turn right onto the N10 and follow this road until the turn-off R72 on your left.
- Follow signs towards Port Alfred and Alexandria.
- Turn off to Woody Cape to the right just before you enter the town of Alexandria.
- Travel the **dirt road** and follow signposts for Woody Cape Section.
- Eventually the entrance to the park offices is on your right.

Connecting with nature

The Woody Cape Section stretches from the **Sundays River Mouth** to the **Bushman's River Mouth** and includes the **Alexandria Dunefield** (one of the largest active dunefields in the world) and the **Alexandria Forest**, also known as **Langebos**, a narrow stretch of pristine indigenous forest bordering the Alexandria Dunefield. This entire area was first managed by Forestry and later by Cape Nature Conservation before it was incorporated into the Addo Elephant National Park in 2002.

Unfortunately, self-drive to this area is limited to the public dirt road from Alexandria to Cape Padrone and eventually Boknes. The turn-off to the Woody Cape Office is on this road. Unless you choose to do a walk or go on one of the hiking trails, you won't see much of the forest or the dunefield from your vehicle.

Alexandria Forest

The Alexandria Forest in the Woody Cape Section is an extraordinary gem; not only because forests are so rare in South Africa (they cover only 0.25% of South Africa's land surface), but rather because this type of forest differs from all other indigenous forests in the country. The Alexandria Forest is unlike the well-known Knysna Forest, the Afro-montane

Southern Coastal Forest

Beginning of hiking trail

forest in the ravines and valleys of the Zuurberg or the coastal dune and rare sand forests of KwaZulu-Natal, but is special because of its own unique species composition. Of the original Alexandria Forest, only 5% is formally conserved at present and most of the rest has vanished, never to re-grow again. Vast areas have been cleared for planted pastures, housing and other human needs.

The structure of the Alexandria Forest is unique because of the reduced light levels beneath the canopy, which results in the relatively low percentage of shade-loving trees. In addition, it houses several endemic tree species that make up about 5% of the total plant types. The upper canopy is almost totally closed and does not reach very high, being between 4–6 m to a maximum of 10 m. Interestingly most of the canopy species have almost no trunks, but branch low down. Therefore, one has the impression of walking through the lower canopy when strolling through the forest. The Alexandria Forest receives the highest rainfall in the AENP, which is about 600–700 mm per year.

Lichen-laden and orchid-laced canopies of the giant **Outeniqua yellow-woods** tower above the rest of the forest. Knysna turacos fly from tree to tree, flashing their red petticoats as they search for fruit. Blazing yellow orioles on blood-red coral tree blooms simply take your breath away with their rare beauty. In the background is the perpetual symphony of birdsong from chorister robin-chats, sombre bulbuls, olive thrushes and collared sunbirds. Other forest species include the black cuckoo (summer only), grey cuckoo-shrike and the spectacular Narina trogon.

Canopy trees include the African white-stinkwood, white-milkwood and the coast coral-tree. Others are the Cape plane, needle bush, white pear, cross-berry, wild caper-bush, common cabbage tree, plumbago, sweet-thorn acacia, scrambling fig and African olive. Some of the true endemics are the Alexandria cycad, bend-me-not and *Atalja capensis.* There is a large population of this rare *Encephalartos capense* in the park. Endemic means the species occurs only in these forests and nowhere else in South Africa or the world.

Between June and October, when the coral trees are in flower and it looks as if red blankets cover parts of the canopy, the beauty of the forest will blow you away.

Alexandria hiking trail

The Alexandria Hiking Trail is a two-day, 36 km hike and is quite demanding (moderate to high level of fitness), but it is the best way to explore and enjoy the forest and vast dunefield.

The hike traverses three different environments, starting in forest before shuffling across the sand dunes and ending with easier strolls on coastal shores beside crumbling cliffs. It's not for the faint-hearted (distances are long, it can get very hot and the sandy terrain makes for hard going in places) but it's an **entirely unique trail within South Africa**. It is also great for birders since more than 250 bird species have been recorded here. The forest has an ancient, magical ambience and the big trees provide good chances of spotting the likes of the Narina trogon, Knysna turaco, trumpeter hornbill and African olive pigeon.

The **Langebos Huts** are the base huts for the trail and it is recommended hikers stay at the huts the night before commencing the trail. The two newly built huts are equipped with six beds each, including mattresses, a fridge, stove, hot water showers, toilets and a braai area.

At the end of the first day, hikers will find the **Woody Cape hut** nestled in thick coastal forest on top of the cliffs with a view of Bird Island in the distance. This hut has 12 beds with mattresses and rainwater tanks. No fires are allowed at the Woody Cape hut and the hikers must carry their own cookers. Water is only for drinking and cooking as the hut relies on rainwater.

At least one or more members of the hiking group should **be able to read a map and compass**, and **wear a brightly coloured jacket** that can be seen from the air, should it be necessary to launch a search and rescue operation. Book at Matyholweni Rest Camp by telephone: +27 (0) 41 468 0916/8 or email: *matyholweni@sanparks.org*.

The Tree Dassie Trail

The short **Tree Dassie Trail** is named after the rare tree dassie (hyrax) that lives in the Alexandria Forest. This arboreal creature is nocturnal and very shy but its shrill call can be heard at dusk.

The trail starts at the office and follows the grass road to **Langevlakte** where it meets up and joins the two-day trail leading back to the office.

Everlasting (*Helichrusum sp.*)

Knysna Turaco (*Tauraco corythaix*)

Chorister Robin-chat (*Cossypha dichroa*)

The Marine Section and offshore islands

Introduction

Addo has both terrestrial and marine components and existing boat charters currently offer boat cruises around the St Croix and Bird Islands if permits allow. No boats are permitted to land on the islands, however, and the park currently generates no income from any such activities. The park also includes the largest coastal transgressive dunefield in the southern hemisphere, which at 88 km in length offers spectacular views and activity potential, an example of which is the existing Alexandria Hiking Trail.

The boundaries of Addo Elephant National Park extend from the Darlington Section north of the Zuurberg Mountain Range, right down to the ocean to incorporate a 120 000 ha **Marine Section**. The latter stretches for 88 km **from Coega to Cannon Rocks** in Algoa Bay and incorporates the Marine Section up to and slightly beyond **Bird Island** and **St Croix Island** groups. AENP took over management of this section on 1 August 2019. This protected area includes both the coastline and marine area where you can see the **largest dunefield in the world** and both **southern right whales** and **great white sharks**. This is why the Addo Elephant National Park can boast the **Big Seven** within its boundaries. There is no other place in the world where you will find elephant, lion, leopard, buffalo, rhino, southern right whale and great white shark within the same protected area.

Two of the Big Seven are not the only extraordinary creatures found here as the park also boasts, among many others, humpback whales, bottle-nose dolphins, seals, the critically endangered African penguins, Cape gannets, roseate and other terns, many migratory and reef fish species and a rich benthic fauna on the reefs.

View of the Marine Section and offshore island

Algoa Bay was recently (18 April 2016) named the **Bottlenose Dolphin Capital** of the world after photographic proof and scientific studies showed that an estimated population of 28 482 individuals make use of these waters, which is the largest along the South African coastline.

The east-coast waters are characterised by the **warm waters** of the southward-flowing **Agulhas Current** that influence the composition of the animal and plant communities along this coastline. The shores support a great diversity of marine organisms, some of which are commercially significant; for example, abalone, mussels and oysters.

The intertidal and the sub-tidal zones provide feeding grounds for many species of fish. Several are important angling species. In the open sea, there is no firm base and organisms must either drift or be able to swim.

Coastal dunefields

The **Alexandria Dunefield** and the **coastal grassy plains** of Addo Elephant National Park stretch between Sundays River Mouth and Bushman's River Mouth. SANParks assumed management of this coastal area in 2002, with its transfer from the provincial conservation department.

Dunefields are evocative **oceans of sand**; moving, rolling and often quite frightening. To give you some perspective on how awesome a display it is, this dunefield is 65 000 years old, part of the largest and least-degraded coastal dunefield in South Africa and not to mention one of the **largest and most spectacular in the world**. The Alexandria Dunefield is a 'must see', even if only from afar.

The Alexandria dunes rise from the beach and occasionally reach a **height of 150 m**. It stretches over **15 800 ha**, with its widest point being **five kilometres from the sea**. These dunes support a unique set of habitats for a range of specialist organisms. On the pristine **sandy beaches** adjacent to the dunefields, threatened bird species, such as the **black oystercatcher**, are breeding. These sandy beaches consist of an unstable sandy bottom layer that is continually modified by waves and currents, resulting in an absence of plants between the tide marks. Yet, a few animal species have adapted to live in this harsh environment. For example, the **plough snail** and **white mussel** have adapted by burrowing in the sand. These animals emerge to feed when conditions are relatively mild, or they sit tight in the sand and filter food particles from the seawater with siphons or strainers.

However, this extraordinary dunefield is **currently under threat** because of the invasive alien plants previously introduced to stabilise the dunes. Unfortunately, private landholdings were not always managed in conjunction with the dunes. On the other hand, the dunefields were managed in isolation from the hinterland. Both the hinterland and the dunes are ecologically interdependent and should be managed as such.

The Alexandria Dunefield is home to many **archaeological sites**, which are the middens of the nomadic **strandloper** (or 'beach walker' translated from Afrikaans) San people. These middens contain shells and bones of animals eaten by the San as well as fragments of pottery and stone implements. In times of drought, early man came to the dunes for water, since they act as an **unconsolidated aquifer;** a process that continues today.

As the Portuguese advanced towards the East, they erected **inscribed limestone crosses** to proclaim their presence. In 1938, Eric Axelson discovered the fragments of the **Kwaaihoek Cross**. Today, the stone copy of the *padrão* positioned by **Bartolomeu Dias in 1488**, on Kwaaihoek, falls within the footprint of the park.

Reed Cormorants (*Microcarbo africanus*)

African Penguins *(Spheniscus demersus)* on Bird Island

Dunes at the Sundays River Mouth

The closest self-drivers can get to the dunefields is by visiting the **Sundays River Mouth**. To get there, enter the **Pearson Park Resort** (private resort) in the Colchester Village and pay a small fee to drive along the Sundays River up to a few hundred metres from the mouth. The view on the low dunes fringing the mouth will give you a good idea of what the dunefields entail. Alternatively, take one of the **whale-watching boat trips** to the islands and along the coast.

Birdwatching here is superb. In the summer months there are large tern roosts that include swifts, sandwich and common terns and the diminutive Damara tern, which breeds in the nearby sand dunes.

Bird Island

The **Bird Island** group is a cluster of low rocky offshore islets, with a lighthouse on the largest one. They are positioned roughly opposite the Woody Cape Section of the park and were first discovered and described by Bartolomeu Dias on his voyage of discovery around the Cape. The largest of the group of three islets, is called **St Croix**, where Dias erected a wooden cross and named it "the island of the holy cross". The coastal islands have impressive breeding colonies of **Cape gannet** and **African penguin** and it is one of the few South African breeding locations for **roseate tern**.

Gannets are among the most beautiful birds in Africa and Bird Island hosts **the largest gannetry** in the world. Gannets feed on sardine as well as anchovy and should these numbers dwindle, it will have a serious effect on the entire food chain. Sadly, this is currently happening since over-fishing has depleted their food source and now these once-prolific populations are seriously endangered and on the IUCN Red List. The population, here at Bird Island and the five other islands around the tip of Africa, has declined by more than half in the last 50 years. The main cause is lack of food. Adult predation by fur seals, oil spills and getting tangled in fishing nets are other causes.

When on land, gannets walk with a rather ungainly waddle, but once airborne they change into the most **graceful flyers**. It is awe-inspiring to watch a flock following a shoal of sardines, diving down into the water to depths of up to 20 m at the speed of torpedoes. **Gannets pair for life** and it remains a miracle how they find their mates among thousands of others, but they always do. They build a nest using guano, to which they add bits of bone, feathers and any debris they can find. Such a pair will produce only **one egg** per year and incubate it using their webbed feet. Both the parents care for the chick, which will take up to 100 days to fledge. While breeding, the foraging parent can fly several hundred kilometres to a food source and back. These are fascinating birds, playing a vital role in the food chain and are worthy of urgent conservation.

St Croix Island

Addo Elephant Park is responsible for the protection of about 50% of the total **African penguin** population, the only penguin species found in Africa. **St Croix Island** near Coega is one of the 28 localities where they breed. They once were South Africa's most abundant seabird, but these quaint birds are currently more threatened than rhino. During the 1920s, estimations indicated about one million breeding pairs. Since then their numbers have decreased alarmingly and in a few years' time, none may be left. The population initially declined until more or less the '60s because of egg collection, guano harvesting and killing of the birds for food. In the meanwhile these activities have been stopped, but since then their numbers steadily declined until 2004, when another population crash of 70% occurred. The large-scale population loss was largely as a result of the lack of available food resources, but also because of oil spillages and competition with commercial fishing. This iconic species is now listed as **Critically Endangered** by the IUCN Red List. Unless they are looked after, our grandchildren will only know them from pictures.

Cape Gannets (Morus capensis) on Bird Island

Greater Double-collared Sunbird *(Cinnyris afer)*

Pale Chanting Goshawk *(Melierax canorus)* and Cape Cobra *(Naja nivea)*

Reptiles

Reptiles do well in semi-arid regions such as the AENP. Since they are ectothermic or **cold-blooded**, they require heat from the sun to warm up and become active. Average maximum day temperatures in the park are usually well above 30 degrees Celsius in summer, which is the optimum temperature at which reptiles generally operate. In winter, when the temperatures drop sharply, reptiles either go into **full hibernation** (examples are the tortoises and monitor lizards), while others only **semi-hibernate**, which means they may come out to bask in the sun but will not really feed. If they do so, it could cause their death because food might rot in the stomach if they have not derived enough energy from the sun to digest it.

Only four species of **tortoise** and **one species of terrapin** are found in the park. The **leopard tortoise** is by far the most common, can grow quite large (30–45 cm) and Domkrag the famous tortoise was one of them. The other three tortoise species are scarce and not often encountered. The **parrot-beaked tortoise** is small (7–10 cm), a relative of the 'padloper', and is rarely seen because it remains under cover to avoid being preyed upon. The **angulate tortoise** is slightly bigger, between 15–25 cm, and is straw coloured with dark edges to the scutes. The **tented tortoise** is also small at 8–12 cm. The only **terrapin** species can be found in the dams and waterholes of the park.

There are three types of **chameleon** – all of them dwarf species. The six **gecko species** are often seen around buildings and so are the **lizard species**, of which there are 14 altogether. **Snakes** are quite at home in Addo but of the 27 species, only a few are poisonous. It is, however, best to leave them alone when you encounter them. Most snakes will slither away long before you notice them, but it is always best to be careful where you go, especially after dark. **Use a flashlight** when walking outside and mind your step. The **puff adder** is one of the most dangerous, mainly because it is lethargic and is often encountered on footpaths where people walk. Other highly venomous snakes such as the Cape cobra, berg adder, night adder, rinkhals, coral snake and boomslang should be avoided. Most of the others are not poisonous or only slightly venomous.

Amphibians are not featured in this book, but it is interesting to know that there are at least 19 species to be found.

Spotted Bush Snake *(Philopthalmus semivarigatus)*

Rock Monitor Lizard (Varanus niloticus niloticus)

Birds

Addo Elephant National Park (AENP) is ecologically one of the most diverse parks in the world. It covers an area that includes five of the seven biomes present in southern Africa. The different habitats created by this versatility are home to a **surprising variety of bird species** – about 418. Quite a number of species are environmentally tolerant and occur in more than one biome, but some have specific habitat requirements and occur only in a particular area.

Although most visitors spend all their time in the main game game-viewing areas, it is good to know that there is also **exciting birdlife** in the other **AENP areas** that could be worth a visit.

The Darlington Section situated to the north of the mountain range is an arid Karoo-like area and it is not surprising to encounter several birds and animals that prefer arid environments and are not found elsewhere in the park. These include birds like **greater kestrels, lesser kestrels, grey-winged francolins, kori bustards, double-banded coursers, white-backed mousebirds, spike-heeled larks, grey-backed sparrow-larks, Karoo chats, Namaqua warblers, rufous-eared warblers, pririt batisses, dusky sunbirds and scaly-feathered finches**.

The large Darlington Dam in this arid section creates an aquatic habitat unlike any other part of the park. Here you can see birds you wouldn't expect in such an otherwise dry area. These include a variety of **waders, flamingoes, terns, cormorants, darters, spoonbills** and, of course, the majestic **African fish eagle**.

The Zuurberg Mountain Range is situated between the Darlington Section and the Main Camp Area. It is an undulating landscape with parallel high ridges and deep valleys.

Pied Starling *(Lamprotornis bicolor)*

Cape Weaver *(Ploceus capensis)*

Southern Red Bishop *(Euplectes orix)*

Ant-eating Chat *(Myrmecocichla formicivora)*

Dark-capped Bulbul *(Pycnonotus tricolor)*

The northern slopes are dry, and further south it becomes gradually wetter. Even fynbos occurs on top of the ridges. Owing to the variety of habitats offered by this rugged terrain you can expect to see some **special localised birds** such as **red-winged francolins, Cape sugarbirds, ground woodpeckers, Verreaux's eagles, marsh owls** and **African scops owls**.

The park includes coastal forests but these are not geographically attached to the Main Game Area. Here you can expect to see typical forest birds such as **African crowned eagles, African olive pigeons, lemon doves, olive woodpeckers, trumpeter hornbills, grey cuckoo-shrikes, chorister robin-chats** and **grey sunbirds**.

The Marine Section of the park, as expected, is quite different from other parts. It includes the estuary of the Sundays River, the coastal region, a very large expanse of dunes and the offshore islands. It offers a huge variety of marine birds that forage on the wing, some of which very seldom, if ever, touch down on solid ground. The beach zone and the estuary host many different types of **waders** and are also home to the endangered **African black oystercatcher**. The offshore islands provide safe breeding areas for the endangered **African penguin** and **Cape gannet**, the largest gannet breeding colony in the world is on Bird Island.

Red Hartebeest *(Alcelaphus buselaphus)* and Spotted Hyena *(Crocuta crocuta)*

Suricate/Meerkat *(Suricata suricatta)*

Mammals

The typical antelope species that occur in Subtropical Thicket include **bushbuck, grey rhebok, mountain reedbuck, eland, greater kudu, red hartebeest, cape grysbok** and **common duiker**.

Large mammals include **elephant, black rhinoceros** and **buffalo** while **hippopotami** are only in the Sundays River in the Kabouga Section and the dam in the Darlington Section. In the Main Game Area **plains zebra** historically occurred and were reintroduced, while the **mountain zebra** appears only in the mountainous parts of the park.

Other mammals include the **primates, vervet monkeys** and **chacma baboons**, the two pig species, **warthog** and **bushpig, hares**, several **bat** species, **rodents** which include the *spring hare* and **porcupine** as well as several **mouse** and **rat** species.

Lions and **spotted hyena** occurred historically and were reintroduced but **leopard, brown hyena, caracal, black-backed jackal** and the termite and ant-eating **bat-eared foxes** were resident when the park was proclaimed. Among the smaller carnivores **suricate** are a special sighting and other mongooses such as **yellow mongoose, grey mongoose** and others occur naturally.

Yellow Mongoose youngsters playing

Enchanting elephants

Elephants are fascinating creatures. Their mere presence commands respect and their enormous size instils quiet awe. Compassion and tenderness towards their young and extended family members give elephants a certain human quality. They never forget and pass down knowledge to their offspring regarding survival strategies, social graces, best behaviours and migratory paths tramped by their ancestors to secure food and water. These animals appear to have wisdom that transcends their animal kingdom. It's as if they know much more than they are telling.

People who have had encounters with elephants across Africa will know that despite the grace and charm of these seemingly docile giants, they can suddenly change when feeling threatened. Within seconds they can become giant packages of fury, trumpeting in shrill tones to evoke terror and flapping their enormous ears to appear even more ominous. In areas where they are persecuted, you seldom encounters matriarchal herds with babies because they mainly visit waterholes under the cover of night. Should you come across a herd during the day, it is best to keep a proper distance.

Addo elephants seem to be different – they are much more relaxed around vehicles, which they ignore and regard as part of the environment – unless people do silly things. **Addo Elephant National Park is the best place in the world to watch and connect with these ever-interesting giants.**

Elephants drink regularly and on hot days like to immerse themselves, roll in the mud, splash and play. Waterholes are therefore the best places in Addo to watch them. They spend most of their day feeding and in the dense thicket they can vanish before your very eyes. Despite there being over 600 individuals in the park, it happens regularly that they don't emerge from the bush and are nowhere to be seen. This usually happens in very cold or extremely windy weather.

At Hapoor Dam, the parking area is in close proximity to the water. To sit safely in your vehicle waiting and watching the first matriarchal herds appear in the distance is exhilarating. The matriarch leads the way and the rest of the herd follow almost in single file as they make their way through the bushes. Once they smell the water and get into the open, their steps quicken and soon some of the youngsters may break ranks, almost overtaking the leader in their haste to quench their thirst.

Few sensations surpass the feeling of awe and wonder as these giants mill around the water right in front of you. It often happens that elephants move out onto the parking area, ignore the vehicles and pass less than a metre from where you are. They have so much confidence in the goodwill of their audience that they will even allow one of their young to follow them.

TREAT ELEPHANT ENCOUNTERS WITH RESPECT

- Drive slowly and always keep a safe distance (50 m) between you and the elephants.
- Avoid splitting a herd and be careful not to block their path when they attempt to cross the road.
- If elephants seem to be comfortable with your presence, switch off your engine, sit quietly and enjoy the sighting.
- Avoid sudden and jerky movements and keep voices down.
- If you notice unease or apprehension, switch engine on, and wait a while before slowly backing away.
- Should the noise of the engine aggravate them, switch off immediately, wait a while for them to calm down, switch on again and slowly retreat.
- Never bang the side of the vehicle, shout, wave, hoot, rev your engine or get out of the vehicle.
- Don't follow too close behind elephants walking in the road ahead – this can stress them.
- If elephants are walking in the road towards you, switch off your engine, relax and don't move, avoid directing any attention to your vehicle, keep voices down and let them pass. At this time, it is too late to apply any avoidance manoeuvres with the vehicle.
- Keep an eye open for bulls in musth as they are unpredictable, especially when they are in the company of a female herd.
- Take note of the tell-tale signs of anxiety in elephants. These include a visible flow from the temporal gland between the eye and the base of the ear, giving the appearance of weeping; 'smelling' with the trunk held high; 'swinging' of front leg; 'crossing' the front legs; 'resting' the trunk over the tusks; bush bashing; and other signs given above.

Understanding elephant body language not only enhances elephant viewing, but can also prevent unfortunate incidents. Treat elephants with respect.

Elephant charges Buffalo

Elephants in Hapoor Dam

There is a close association between Elephants and the Flightless Dung Beetle

Conservation challenges for the park

An important part of the Addo Elephant National Park's mission is to conserve the integrity and diversity of nature. This includes all important and critically endangered or threatened species, of which there are several in the park. One of the critically endangered mammals is the **black rhino**. Other mammals at risk include the elephant, leopard and lion. The **flightless dung beetle** (associated with the dung of elephant and of other mammals that occur in the thicket) is an **endemic** insect that has been threatened with extinction and is now classified as 'vulnerable' in the Red Data Book of Threatened Species.

There are many more animal and plant species that need special protection, but these will not be discussed here. Some of the endangered mammals are **keystone species**, which have a huge impact on the health of the ecosystem. These include marine species such as whales and sharks.

The **keystone concept** comes from the age-old craft of stone-masonry. In a stone arch, the keystone in the centre plays a vital role. It sustains the least pressure of all the stones that make up the arch but without it, the arch will collapse. In nature the same principle applies. There are certain keystone species that play a critical role in the sensitive relationship between wildlife and the ecosystem. They help maintain the ecological framework and the entire natural community. Without them the balance in nature will collapse. The black rhino, the elephant and the large predators – both on land and in the Marine Section – are keystone species in the AENP and there would be serious consequences if they were removed from the ecosystem.

Black rhinoceros conservation

Rhinos are large herbivores, typically weighing more than 1 000 kg, and consume very large amounts of plant material. Their pivotal role as a keystone species is **maintaining the Subtropical Thicket vegetation**. Subtropical Thicket supports many animal species, either directly in browsers themselves, or indirectly in predators that prey on them. By preferring to eat certain plants, black rhinos **increase biodiversity** by giving other plant species a better chance and space to grow. Thicket species are a constant source of sustenance, building nutrients by absorbing carbon dioxide and manufacturing sugars and starches by photosynthesis.

South Africa has the largest population of rhino in the world but in the past decade they have become one of the favourite targets of poachers. According to statistics released by the Department of Environmental Affairs, in the 10-year period to August 2018, a total of 7 638 rhino were poached in South Africa. Estimates suggest fewer than 20 000 white rhino and 2 000 black rhino now remain in the country. The gravity of this situation is illustrated when looking at the history of rhino numbers over the previous century. Throughout most of the 20th century, the **black rhino boasted the largest population numbers of all the world's rhinos**, with estimates of about 850 000 individuals. However, by 1960, only about 100 000 remained, owing to land clearances for human settlement and farming, large-scale poaching and persistent hunting of the species. **Between 1960 and 1995, the population decreased by a dramatic 95.6%**. While numbers are steadily on the increase today, current estimates are still 90% lower than three generations ago, with one of the four subspecies now considered extinct.

Black Rhinoceros

RHINO HORN: MYTHS AND TRUTHS

Myth: Rhino horn has medical properties
In a number of Eastern countries, rhino horn is believed to cure a variety of ailments. In traditional Chinese medicine, the horn, which is shaved or ground into a powder and then dissolved in boiling water, is used to treat conditions such as fever, rheumatism and gout. It is also believed to cure snakebites, hallucinations, typhoid, headaches, carbuncles, vomiting, food poisoning and 'devil possession'. There is no evidence to back any of these claims.

Myth: Rhino horn is an aphrodisiac
Some Asian cultures believe that rhino horn has magical power to increase libido. This belief is totally unfounded, as demonstrated by several scientific studies. Why kill a rhino and pay millions, rather than buy some of the many pharmaceutical products that really do work and are available online?

Myth: Rhino horn is a party drug
Rhino horn is apparently the new drug of status in Vietnam and is more expensive than cocaine. The users grind it into a powder, which is mixed with water or wine for consumption. They believe that drinking a tonic made from the horn will detoxify the body after a night of heavy drinking, and prevent a hangover. None of this is true, and for a rhino to be killed to avoid a hangover beggars belief.

Myth: Rhino horn defines status
Some cultures value the rhino horn as 'bling'. The horns are fashioned into everything from handles to accessories, which are deemed status symbols. There should be laws to criminalise this behaviour and people who value the product for spurious reasons should be ostracised by society. This was successfully achieved with the trade in ivory and fur, and rhino horn should be added to the list. But enforcement must be international and effective.

Truth: Rhino horn is made of keratin
The horn of a rhino is composed of keratin, the same material as human fingernails and hair. We may as well grind finger- or toenail clippings to a powder, add water and drink it. This is how ridiculous the belief is that rhino horn has special properties. Why spend millions for something that is just another form of keratin?

Truth: China has partially lifted the ban on the trade in rhino horns
Conservationists are enraged by this unexpected move and say it will further jeopardise already imperilled species. This could push rhinos into extinction by emboldening traffickers to poach and stockpile goods for a country that places exceptional value on animal parts.

Truth: There is a war out there
The war is fought by ruthless criminals who plunder our game reserves to enrich themselves. They are determined, fearless, well trained and will kill endangered animals without a second thought. Like all wars, protecting what is ours is an expensive business.

Truth: Our wildlife heritage is at stake
There has been a drastic increase in the number of rhino poaching incidents in national parks. At the present rate of poaching, rhinos will be extinct in our lifetime. Extinction is forever. Our heritage will be gone.

Additional information

Accommodation in the park and in concession areas

SANPark accommodation

Addo Elephant National Park offers a range of accommodation types. For comprehensive information on these, including photographs of the facilities and more detail on how the units are equipped, please visit the website *https://www.sanparks.org/parks/addo/tourism/accommodation.php*

Gorah Elephant Camp

Situated in the Main Game Area section of the park – a five-star facility with luxury tented camps.
Visit the website: *www.gorah.hunterhotels.com*
Tel: +27 (0) 44 501 1111.

River Bend Lodge

Situated in the Nyathi area of the park, it offers luxury five-star accommodation in eight rooms as well as a Private Safari Villa accommodating six guests. There is a Wellness Centre and children of all ages are welcome. Visit the website: *www.riverbendlodge.co.za* or contact them on +27 (0) 42 233 8000.

Pool at Main Camp

Important contact information

For enquires, contact Addo Elephant National Park
Tel: +27 (0) 42 233 8600 (during office hours – 07:00–19:00)
Fax: +27 (0) 42 233 8643 (during office hours – 07:00–19:00)
Email: *addoenquiries@sanparks.org*

- For emergencies, contact the Hospitality Manager after hours on +27 (0) 82 471 0267.
- The nearest hospital and doctor are in Kirkwood, 35 km from the park.
- Hospital tel (after hours) +27 (0) 42 230 0406.
- Doctor tel (office hours) + 27 (0) 42 230 1082.
- The nearest private hospital is in Gqeberha, 75 km from the park.
- Greenacres Hospital tel +27 (0) 41 390 7000.
- The nearest police station is in Addo, 15 km from the park: tel +27 (0) 42 233 0314.

General information as given by the SANParks website
https://www.sanparks.org/parks/addo/tourism/general.php

Climate

The climate is temperate to hot and temperatures in summer (November to February) may reach over 40 degrees Celsius. Winter days are mild but cold at night with frost at times. The average annual rainfall is 450 mm and is spread throughout the year, although peaks do often occur in February–March and October–November.

Day visitors

- No booking is required for a day visit, unless the visitor wants to participate in organised activities.
- Day visitors can participate in a wide variety of activities, including wildlife viewing from their own vehicle, guided game drives, horse rides, hiking trails, the bird hide, the PPC Discovery Trail, picnic and braai sites, the shop and restaurant.
- Day trips can be made to the outlying sections of the park: the Zuurberg Mountains, Kabouga, Darlington Dam and Woody Cape.

Office and gate hours

- Main Camp Reception is open from 07:00–18:00 daily.
- Matyholweni reception is open from 07:00–17:00 daily, while the gate is open until 18:00.
- The gates are staffed by security guards after closing time. Visitors with accommodation bookings can still enter and exit the gate after closing times provided they inform Reception of their intentions.

Check-in and check-out times

- Check-in time is 14:00 for all accommodation.
- Check-out time is 10:00 for all accommodation units. Check-out time for caravan and tent sites is 12:00.

Park regulations

- To ensure a safe and happy trip through our parks, kindly adhere to the **Rules and Regulations** as stipulated by South African National Parks.
- The use of drones inside (and over) our national parks is strictly prohibited.

Tips and hints

- The swimming pool is for overnight guests only.
- Hot days at waterholes are best for elephant viewing.
- Lions and spotted hyenas are most often seen in the early morning and evening, or at night.
- Pets are not allowed in national parks.
- Fuel for vehicles is available in all parks (or on the park periphery).
- Firearms are to be declared at reception where they will be sealed. The seal will be broken upon departure.

What visitors need to take along in preparation for the trip

- Binoculars, cameras, walking shoes and wildlife reference books.
- A torch or head torch/headlamp is required when walking outside at night and also because outdoor lighting in camps is limited.
- A road map of tourist routes in the park will be given to each vehicle upon entering the park. Additional maps are for sale at the shop. Various information sheets are available at Reception.
- For summer (September–March), cool comfortable clothing is recommended along with sunhats and sunscreen lotion.
- For autumn and winter (April–August), warm clothing is recommended, especially for the nights.
- Insect repellent is recommended, especially in summer.
- No immunisations are needed for travel to the park as it is situated in a malaria-free area.

Elephants in a greeting ritual

Selected bibliography

- Apps, Peter. 1979. *Wild Ways. Field Guide to the Behaviour of Southern African Mammals.* Southern Book Publishers (Pty) Ltd.
- Branch, Bill. 1988. *Field Guide to the Snakes and other Reptiles of Southern Africa.* Cape Town, South Africa: Struik.
- Carnaby, Trevor. 2006. *Beat About the Bush. Mammals.* Jacana Media.
- Carruthers, Vincent. 2000. *The Wildlife of Southern Africa. A Field Guide to the Animals and Plants of the Region.* Struik Publishers.
- Chittenden, Hugh. 2012. *Roberts Bird Guide. A Comprehensive Field Guide to Over 950 Bird Species in Southern Africa.* The Trustees of John Voelcker Bird Book Fund.
- Derichs, Peter. 2015. *Peter's Guide. Addo Elephant National Park (Main Camp).* Peter Derichs, Honeydew.
- Elephant Specialists Advisory Group. 2017. *Understanding Elephants. Guidelines for safe and enjoyable elephant viewing.* Struik Nature, an imprint of Penguin Random House South Africa (Pty) Ltd. Cape Town.
- Eltringham, Dr S K. 1991. *The illustrated encyclopedia of Elephants, From Their Origins and Evolution to their Ceremonial and Working Relationship with Man.* Salamander Books Limited, London, New York.
- Engelbrecht, Henriette *et. al.* 2019. *South African National Parks Official Guide: Addo Elephant National Park.* Tourism Blueprint.
- Estes, Richard D. 1993. *The Safari Companion.* Russel Friedman Books.
- Hockey, PAR, Dean, WRJ & PG Ryan. *Roberts Birds of Southern Africa VIIth Edition.* The Trustees of John Voelcker Bird Book Fund.
- Manning, John. 2009. *Field Guide to Wild Flowers of South Africa.* Struik Nature, an imprint of Penguin Random House South Africa (Pty) Ltd. Cape Town.
- Mills, Gus and Hes, Lex. 1997. *The Complete Book of Southern African Mammals.* Struik Winchester.
- Pinnock, Don & Bell, Colin. 2019. *The Last Elephants.* Struik Nature, an imprint of Penguin Random House South Africa (Pty) Ltd. Cape Town.
- Pooley, Elsa. 1998. *A Field Guide to Wild Flowers: KwaZulu-Natal and the Eastern Region.* Natal Flora Publications Trust.
- Skinner, John D & Chimimba, Christian T. Revised 2005. *The Mammals of the Southern African Sub-region.* Cambridge University Press.
- Smith, Gideon F & Crouch Neil R. 2009. *Guide to Succulents of Southern Africa.* Struik Nature, an imprint of Penguin Random House South Africa (Pty) Ltd. Cape Town.
- South African National Parks. *Information for Guides Operating in the Addo Elephant National Park.* Pdf.
- Stuart, Chris & Tilde. 1997. *Field Guide to the Mammals of Southern Africa.* Struik Publishers, Cape Town.
- Urquhart, Colin & Klages, Norbert. 1997. *Addo, More Than Just Elephants – Your Guide to the Addo Elephant National Park South Africa.* Bluecliff Publishing.
- Mucina, L. & Rutherford, M.C. (eds.) 2006. *The vegetation of South Africa, Lesotho and Swaziland.* Strelitzia 19. South African National Biodiversity Institute, Pretoria.

Blackbacked Jackal and Pied Crow scavenging

Male and female Lion *(Panthera leo)*

Behind the scenes

We find elephants fascinating, and ever since we started our regular visits to the Addo Elephant National Park, we have enjoyed the privilege of watching and marvelling at these giant beasts. Initially, however, on our first visit in 1983, we didn't see a single elephant during our stay. This was because of the opening of the Gorah farm that was overgrown with prickly pears, which no elephant can resist. We didn't know this and attributed their absence to the gale-force winds that were blowing. Our disappointment was profound, but fortunately this did not deter us from paying regular visits again and again.

In the early years the road network was limited to Domkrag Dam, the Nzipondo, Zuurkop and Hapoor loops. At places the roads were mere tracks that became slippery messes after the slightest shower. We remember several occasions where we helped recover vehicles that were stuck in the muddy porridge. The upgrading and tarring of the main roads were some of the huge improvements. With each visit we noticed progress at all levels. New land was acquired, new sections opened, accommodation upgraded, a new camp and entrance gate developed and a marine section included. And all this is not yet the end of the dream. More is to follow.

In 2002, when HPH Publishing was still in its infancy, we produced our first book on the park – a slim booklet on what Addo Elephant National Park offered – and called it ***Greater Addo, Route of Giants***. At the time, many of the developments we now may take for granted were still mere dreams for the future. We are filled with admiration at what the South African National Parks, with the help of its generous donors, have achieved in the past 20 years and wish to congratulate the management and staff of the Addo Elephant National Park, past and present, on a task well done.

With the present book, ***Addo Elephant National Park Self-drive, Routes, Roads and Ratings***, the fourth in the series on our favourite parks, we hope to share knowledge and photographs of the beautiful flora and fauna to be seen during a visit to the park. They have created something South Africa can be proud of.

HPH Publishing is a family enterprise specialising in books on the natural world and related topics. Heinrich van den Berg plays a pivotal role in the team. He lives in Johannesburg and runs HPH Publishing. Both Philip and Heinrich are the principal photographers, while I contribute to a lesser extent. The writing falls mainly on my shoulders, although Heinrich contributes to the text with his creative insight.

The best part of making a book is when everything comes together. This is when we sit down to select the images and build the book. Eventually, when all that is done, the files are ready to be sent to the printers. Developing a book requires a lot of different proficiencies. Heinrich's degree in civil engineering has given him many of the skills he uses every day to run a thriving business, while his creative flair contributes to his success as a professional wildlife photographer, film maker, creative writer and designer. With his background of environmental education at the highest level, Philip has a lifetime of experience in interpreting nature, and his holistic approach and extensive knowledge is invaluable. My post-graduate training in zoology, and proficiency in botany have enabled me to tackle the text with confidence.

We wish the Addo Elephant National Park all the best for the future. May it grow and thrive so that there will always be a place where wildlife roams free in a balanced natural environment, and where we and generations to come can visit to reconnect with nature.

Ingrid van den Berg

Acknowledgements

The Van den Berg team thanks the manager and staff of the Addo Elephant National Park and all the other contributors who helped to get this book on to the shelves in the very difficult circumstances of the 2020 Covid-19 pandemic. In particular, we thank Nick de Goede (Park Manager) for writing the engaging foreword, his kind assistance and his enthusiasm for the project; the Regional Ecologist of the South African National Parks, Charlene Bissett, who made a huge contribution in checking the manuscript for factual correctness. She was assisted by Cathy Dreyer (Conservation Manager at the time of writing) and the Section Rangers Anban Padyachee, Ilse Welgemoed, Jessyca Domingo, Johannes Jaantjies, Michael Paxton and Rob Milne.

Photographic credits: Karin Mitten p.116.
Gerhard Vosloo p.56, p.100 and p.101

Yellow mongoose *(Atilax paludinosus)*

First Edition
ISBN 978-1-77632-320-3
Text by Ingrid van den Berg
Photography by Philip and Ingrid van den Berg, Jean du Plessis, Heinrich van den Berg, Gerhard Vosloo and Karin Mitten
Publisher: Heinrich van den Berg
Edited by Ronél Bouwer
Proofread by Margy Gibson
Design, typesetting and reproduction by Heinrich van den Berg and Nicky Wenhold
Maps by Linda McKenzie – Digital Earth, GIS Consulting and Nicky Wenhold
Printed and bound in China

First edition, first impression 2022
Published by **HPH Publishing**
50A Sixth Street, Linden, Johannesburg, 2195, South Africa
www.hphpublishing.co.za
info@hphpublishing.co.za